The PHYSICAL WORLD

DESCRIBING MOTION

IoP
Institute of Physics Publishing
Bristol and Philadelphia
in association with
The Open University

The Physical World Course Team

Course Team Chair	Robert Lambourne
Academic Editors	John Bolton, Alan Durrant, Robert Lambourne, Joy Manners, Andrew Norton
Authors	David Broadhurst, Derek Capper, Dan Dubin, Tony Evans, Ian Halliday, Carole Haswell, Keith Higgins, Keith Hodgkinson, Mark Jones, Sally Jordan, Ray Mackintosh, David Martin, John Perring, Michael de Podesta, Ian Saunders, Richard Skelding, Tony Sudbery, Stan Zochowski
Consultants	Alan Cayless, Melvin Davies, Graham Farmelo, Stuart Freake, Gloria Medina, Kerry Parker, Alice Peasgood, Graham Read, Russell Stannard, Chris Wigglesworth
Course Managers	Gillian Knight, Michael Watkins
Course Secretaries	Tracey Moore, Tracey Woodcraft
BBC	Deborah Cohen, Tessa Coombs, Steve Evanson, Lisa Hinton, Michael Peet, Jane Roberts
Editors	Gerry Bearman, Rebecca Graham, Ian Nuttall, Peter Twomey
Graphic Designers	Steve Best, Sue Dobson, Sarah Hofton, Pam Owen
Centre for Educational Software staff	Geoff Austin, Andrew Bertie, Canan Blake, Jane Bromley, Philip Butcher, Chris Denham, Nicky Heath, Will Rawes, Jon Rosewell, Andy Sutton, Fiona Thomson, Rufus Wondre
Course Assessor	Roger Blin-Stoyle
Picture Researcher	Lydia K. Eaton

The Course Team wishes to thank Derek Capper for his very substantial contribution to this book. The book made use of material originally prepared for the S271 Course Team by John Bolton and Milo Shott. For the multimedia packages *Functions and derivatives* and *Simple harmonic motion*, thanks are due to Jane Bromley, Alan Cayless, Gloria Medina and Jon Rosewell.

The Open University, Walton Hall, Milton Keynes MK7 6AA

First published 2000

Written, edited, designed and typeset by the Open University.

Published by Institute of Physics Publishing, wholly owned by The Institute of Physics, London. IoP Publishing, Dirac House, Temple Back, Bristol BS1 6BE, UK.

US Office: Institute of Physics Publishing, The Public Ledger Building, Suite 1035, 150 South Independence Mall West, Philadelphia, PA 19106, USA.

Printed and bound in the United Kingdom by the Alden Group, Oxford.

ISBN 0 7503 0715 3

Library of Congress Cataloging-in-Publication Data are available.

This text forms part of an Open University course, S207 *The Physical World*. The complete list of texts that make up this course can be found on the back cover. Details of this and other Open University courses can be obtained from the Course Reservations Centre, PO Box 724, The Open University, Milton Keynes MK7 6ZS, United Kingdom: tel. +44 (0) 1908 653231; e-mail ces-gen@open.ac.uk

Alternatively, you may visit the Open University website at http://www.open.ac.uk where you can learn more about the wide range of courses and packs offered at all levels by the Open University.

To purchase this publication or other components of Open University courses, contact Open University Worldwide Ltd, The Berrill Building, Walton Hall, Milton Keynes MK7 6AA, United Kingdom: tel. +44 (0) 1908 858785, fax +44 (0) 1908 858787, e-mail ouwenq@open.ac.uk; website http://www.ouw.co.uk

1.1

s207book2i1.1

DESCRIBING MOTION

Introduction

Motion is vital to life, and to science. In many ways it was the investigation of motion, initiated by Galileo Galilei in the late sixteenth century, and brought to a head by Isaac Newton in the seventeenth, that inaugurated the modern era of physics. Progress since that time has been so great that describing motion is now regarded as a fundamental part of science rather than one of its frontiers. Nonetheless, the description of motion played a central role in Einstein's formulation of the special theory of relativity in 1905, and it continues to provide an excellent starting point for the quantitative investigation of nature.

The concepts that have been developed to allow the description of motion — concepts such as *speed*, *velocity* and *acceleration* — are now so much a part of everyday language that we rarely think about them. Just consider the number of times each day you have to describe some aspect of motion or understand an instruction about motion; obey a speed limit or work out a journey time. We may take the description of motion for granted, but the concepts involved are so fundamental and so much depends upon them that they really deserve careful consideration. This was clearly understood by Einstein, but it was also well known long before his time.

To the ancient Greek philosopher Zeno, motion seemed such a self-contradictory feature of the world that he and his followers became convinced that the apparent existence of motion only served to indicate the fundamental unreliability of our senses. For Zeno the description of motion was not only a fundamental problem, it was, perhaps, *the* fundamental problem.

Figure 0.1 The paradox of Achilles and the tortoise. Achilles moves faster than the tortoise, but each time he reaches the previous position of the tortoise the tortoise has moved on. The gap gets smaller and smaller but, according to Zeno, never completely vanishes. Does this really mean that Achilles can never quite catch up with the tortoise? Such a conclusion would clearly conflict with our everyday experience of how things move.

Zeno's arguments against the reality of motion are still remembered because of three paradoxes he based upon them. The best known of these paradoxes concerns a race between the ancient hero Achilles and a tortoise (Figure 0.1). Naturally, Achilles can run much faster than a tortoise, so the outcome of the race seems obvious. Even if the tortoise is given a head start, Achilles will quickly overtake it and go on to win the race. However, Zeno argued, this cannot really be the case. According to Zeno, whatever head start is given to the tortoise, Achilles will take some time to reach the starting position of the tortoise and during that time the tortoise, no matter how slowly it moves, will have reached some new position, still ahead of Achilles. Now, starting from the tortoise's original position, Achilles will take a much shorter time to reach the new position of the tortoise, but by the time he does so the tortoise will again have moved on a little, so the reptile will still be ahead of the athlete. In Zeno's view this process can go on forever with the tortoise always moving on, at least a little, in the time that Achilles takes to reach its previous position. Achilles, according to Zeno, will never quite manage to close the gap. Since this conclusion disagrees with everyday experience, Zeno concluded that everyday experience was misleading. Unlike most modern scientists, Zeno preferred to trust his reason rather than his experience of the world.

Modern science is able to resolve Zeno's paradox. Motion is not in conflict with reason, but, as you will see in Chapter 1 of this book, the resolution relies on mathematical concepts that were not known to the ancient Greeks, nor even to Galileo. Chapter 1 deals with motion along a line and with the ways in which such motion can be represented. It will show you how *graphs* can be used to depict motion and how *equations* can provide even more powerful summaries of such graphical information. Crucially, it will also introduce you to some of the basic ideas of *differential calculus*, the branch of mathematics that concerns small changes and their influence. In Chapter 2 these ideas will be extended to situations involving motion in a plane or in space. This will involve another mathematical development — the introduction of *vectors*. In Chapter 3 you will be able to gain experience in using both the calculus notation and vectors as you investigate a variety of *periodic motions* i.e. motions that regularly repeat themselves. These periodic motions include such important cases as the orbital motion of planets and the oscillatory motion of vibrating machinery. Here again, the study of motion requires the use of mathematics — in this case the use of *trigonometric functions* (mainly sine and cosine) that also have a repetitive or periodic character. These three chapters, together with the consolidation and skills development material of Chapter 4, have been carefully designed to teach you a great deal about the scientific process of describing motion, but they have also been designed to equip you with general mathematical and scientific skills that will be of use throughout your subsequent studies.

Open University students should leave the text at this point and view Video 2 *Newton's Revolution*. You should return to the text when you have finished viewing the video.

Chapter 1 Motion along a line

1 From drop-towers to *Oblivion* — some applications of linear motion

We have all experienced that momentary feeling of lightness when an elevator begins its downward motion. It is almost as if our weight had suddenly been reduced or, conceivably, that the pull of the Earth's gravity had decreased for a moment. But imagine what it would be like if the lift cable had suddenly snapped and the lift, with you in it, had plummeted downward. Apart from stark terror, what else do you think you would experience during your fall? What would the *physical experience* of such a disaster be like?

Well, it would be just like jumping from a high tower. If your descent was unimpeded by the resistance of the air, almost all sense of weight would vanish while you were falling. You would feel weightless, just as though you were an astronaut in outer space.

Not surprisingly, scientists who want to know how equipment will behave under the conditions found in spacecraft are keen to simulate the same conditions here on Earth. One way in which they can do this is by dropping their equipment from the top of a tower, or down a vertical shaft. There are a number of research centres around the world where drop facilities of this kind are available. These are specialized facilities where steps are taken to avoid or overcome the effects of air resistance: simply dropping an object in the Earth's atmosphere is not a satisfactory way of simulating the environment of outer space.

Figure 1.1 The 140 m drop-tower in Bremen, Germany.

Figure 1.1 shows the 140 m drop-tower in Bremen, Germany. The tower is airtight, so all the air can be pumped out. Equipment under test is placed inside a specially constructed drop-vehicle and monitored by closed-circuit TV as it falls from the top to the bottom of the tower. About five seconds of free fall can be achieved in this way. During those few seconds, within the falling drop-vehicle, the effects of gravity are reduced to a tiny fraction of their usual value, a condition known as 'microgravity'.

Figure 1.2 The linear motion of a falling test vehicle is stopped safely at the NASA Lewis Zero Gravity Research Facility in Ohio, USA.

In the USA, at the Lewis Research Center in Ohio, NASA operates a 143 m drop-shaft, as part of its Zero Gravity Research Facility. Microgravity investigations conducted at the research facility have concerned the spread of fire, the flow of liquids, and the feasibility of space-based industrial processes that would be impossible under normal terrestrial conditions. Figure 1.2 shows the facility's bullet-shaped drop-vehicle being given a soft landing at the end of a drop, to avoid destroying the expensive equipment that it contains.

At the time of writing, the world's longest drop-shaft is in Japan. The Japan Microgravity Center (JAMIC) has a 700 m drop housed in a disused mine shaft. It would be impossible to evacuate the air from such a big shaft, so in this case the rocket-shaped test capsule is propelled down the shaft by gas-jets with a thrust that is designed to compensate for air resistance. Inside this capsule, there is a second capsule and the space between the capsules is a vacuum. The experiments are carried out in the inner capsule which, to a very good approximation, is in free fall. The two capsules decelerate during the final 200 m of the fall.

By the time you finish this chapter you should be able to work out the duration of the fall in the JAMIC facility, and the highest speed attained by the capsule. You should also be able to work out the length of shaft that would be required to produce any given duration of microgravity.

If all this sounds a bit esoteric you might prefer to consider a different kind of drop-facility. Figure 1.3 shows *Oblivion*, a ride at the Alton Towers Adventure Park, UK. *Oblivion* is described as 'the world's first vertical-drop roller-coaster'. It will not simulate the space environment, but it will produce a few seconds of terror from a simple application of linear motion.

Figure 1.3 *Oblivion*, the vertical-drop roller-coaster at the Alton Towers Adventure Park, UK.

2 Positions along a line

2.1 Simplification and modelling

Everyday experience teaches us that unconfined objects are free to move in three independent directions. I can move my hand up or down, left or right, backwards or forwards. By combining movements in these three directions I can, at least in principle, move my hand to any point in space. The fact that there are just three independent directions, and that these suffice to reach any point, shows that the space in which my hand moves is **three-dimensional**.

The motion of a large object, such as an aeroplane, moving in three-dimensional space is very difficult to describe exactly. The aeroplane may flex, rotate and vibrate as it moves, and there may be complicated changes taking place within it. To avoid such complexities at the start of our investigation of motion we shall initially restrict our attention to objects that move in just one dimension along a line.

We shall treat the object concerned as a **particle**, that is, a point-like concentration of matter that has no size, no shape and no internal structure.

Treating a real object, such as an aeroplane, as though it is a particle is clearly a simplification. Real objects certainly do have size, shape and internal structure, but such details can often be neglected in specific contexts. Making simplifications of this kind is an important part of the skill of scientific modelling in physics. A good **model** uses the well-defined concepts of physics to represent the essential features of a problem while omitting the irrelevant details. The trick is not to oversimplify. The model should be as simple as it can be, but no simpler. Just what this entails will depend on the problem being analysed. For example, the use of the year as a unit of time is a result of the orbital motion of the Earth around the Sun. This orbital motion is described quite easily while treating the Earth as a particle. The Earth's diameter is about 10 000 times smaller than the distance between the Earth and the Sun, so a particle model is a very good approximation in this case. However, a particle model of the Earth cannot account for the distinction between day and night since that depends on the rotation of the Earth.

In this chapter we shall only consider problems that can be adequately modelled by particles moving in one dimension, that is, along a straight line.

Describing the motion of a particle moving along a line may sound like a fairly simple undertaking, but, as you will see, it will present plenty of challenges and will allow us to gain significant insights into the operation of systems such as zero gravity drop-towers and vertical-drop roller-coasters.

Question 1.1 List some more examples of real motions that might, in your opinion, be reasonably well modelled by particles moving along a line. ■

The branch of physics that is concerned with the description of motion is known as **kinematics**. Kinematics is not concerned with *forces*, nor with the causes of motion; those topics are central to the study of *dynamics*. Typical questions that we might ask about the kinematics of a particle are:

- Where is the particle?
- How fast is it moving, and in what direction?
- How rapidly is it speeding up or slowing down?

How such questions are to be answered is the main concern of the rest of this chapter.

2.2 Describing positions along a line

To take a definite case, consider a car moving along a straight horizontal road. The car can be modelled as a particle by supposing the particle to be located at, say, the midpoint of the car. It is clearly convenient to measure the progress of the car with respect to the road, and for this purpose you might use the set of uniformly spaced red-topped posts along the right-hand side of the road (see Figure 1.4). The posts provide a way of assigning a unique **position coordinate** to the car (regarded as a particle) at any instant. The instantaneous position coordinate of the car is simply the number of the nearest post at that moment.

Figure 1.4 A long straight road and a set of uniformly spaced posts along it.

When used in this way the posts provide the basis of a one-dimensional **coordinate system** — a systematic means of assigning position coordinates along a line. Taken together, the posts constitute an **axis** of the system; a straight line along which distances can be measured. One point on the axis must be chosen as the **origin** and assigned the value 0. Points on one side of the origin can then be labelled by their distance from the origin (10 m or 20 m say), while points on the other side are labelled by *minus* their distance from the origin (−10 m or −20 m for example). Conventionally, we might represent any of these values by the algebraic symbol x, in which case it would be called the x-coordinate of the corresponding point on the axis, and the axis itself would be called the x-axis.

Figure 1.5 shows an example, corresponding to a particular choice of origin, and with x increasing smoothly from left to right, as indicated by the arrow. Note that the numbers x increase from left to right everywhere on the x-axis, not only on the positive segment. Thus for example −10 is larger than −20, and so if you subtract −20 from −10 you get $-10 - (-20) = 10$, a positive number. At the time illustrated the car is at $x = 30\,\mathrm{m}$, and a pedestrian (also modelled as a particle) is at $x = -20\,\mathrm{m}$. Notice that it is essential to include the units in the specification of x. It makes no sense to refer to the position of the car as being '30', with no mention of the units, i.e. metres. In one dimension, the specification of a physical **position** x consists of a positive or negative number multiplied by an appropriate unit, in this case the metre.

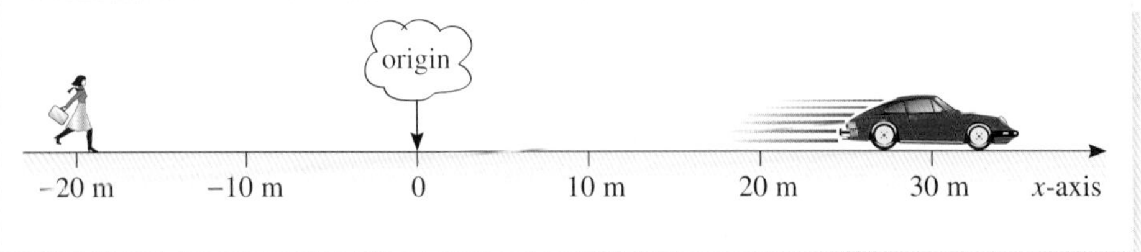

Figure 1.5 A choice of x-axis. Setting up the axis involves choosing an origin and a direction of increasing x (from left to right). Distances are measured in metres (m).

It is worth emphasizing that setting up an x-axis involves some degree of choice. The origin and the direction of increasing x are both chosen in an arbitrary way, usually so as to simplify the problem. The choice of units of measurement is also arbitrary, though usually guided by convention. Even the decision to call the axis an x-axis is arbitrary; x is conventional, but not compulsory. It does not really matter *what* choices are made, but it is essential to stick to the *same* choices throughout the description of a given motion.

The conventions about units deserve special attention, so they have been set apart from the main text in Box 1.1.

Box 1.1 Introducing SI units

The standard way for scientists to measure lengths, distances and positions is in metric units (metres, millimetres, kilometres, etc., as opposed to miles, feet or inches). This is part of an internationally agreed system of units known as **SI** (which stands for the *Système Internationale*). This course uses SI units throughout.

The standard SI abbreviation for the **metre** is m. Since 1983 the speed of light travelling in a vacuum has been defined to be exactly

$$2.997\,924\,58 \times 10^8 \text{ metres per second.}$$

This is an internationally agreed *definition*, not the result of a measurement, so the metre may be similarly defined as the distance that light travels through a vacuum in $\frac{1}{299\,792\,458}$ second. Typical lengths of interest to physicists range from the diameter of the atomic nucleus, which is about 10^{-15} m, to the diameter of the visible Universe, which is about 10^{27} m.

In view of the wide range of lengths that are of interest, it would be inconvenient to use only the metre for their measurement. To avoid this there are standard **SI multiples** and **SI submultiples** that may also be used. You will already be familiar with some of these; the prefix **kilo** means 10^3 as in kilometre and the prefix **milli** stands for 10^{-3} as in millimetre. Table 1.1 gives the standard SI multiples; you are not expected to remember all of them but it is certainly worth learning the more common ones.

Another example of an SI unit is the **second** which is the unit of time. The standard abbreviation for the second is s. (Notice that, like m, *this* symbol is always lower case. The distinction *does* matter.) The time values of interest to experimental physicists range from the 10^{-24} s duration of certain events in subatomic physics, to the present age of the Universe, which is about 10^{18} s, though speculations about the birth and death of the Universe have involved times ranging from 10^{-45} s to 10^{140} s, or more.

Table 1.1 Standard SI multiples and submultiples.

Multiple	Prefix	Symbol for prefix	Submultiple	Prefix	Symbol for prefix
10^{12}	tera	T	10^{-3}	milli	m
10^9	giga	G	10^{-6}	micro	μ
10^6	mega	M	10^{-9}	nano	n
10^3	kilo	k	10^{-12}	pico	p
10^0			10^{-15}	femto	f

Having chosen a convenient x-axis and selected units of length and time, we are well placed to describe the position of a particle as it moves along a straight line. All we need to do is to equip ourselves with a clock, choose an origin of time (that is, an instant at which the time $t = 0$ s) and note the position of the particle at a series of closely spaced intervals. Table 1.2 shows typical results for the kind of car in Figure 1.5. The use of an oblique slash (/) or *solidus* in the column headings in Table 1.2 is another convention. It reminds us that the quantity to the left of the solidus is being measured in the units listed on the right of the solidus.

In fact, if you remember that a physical quantity such as x represents the *product* of a number and a unit of measurement, as in 20 m, for example, you can see that x/m may be thought of as indicating 'the value of x divided by 1 metre', which would just be the number 20 if x was 20 m. It therefore makes good sense to see that the entries in the x/m column of Table 1.2 are indeed just numbers. In effect, x/m indicates that the units have been divided out. Take care always to remember that symbols such as x, that are used to represent physical quantities, conventionally include the relevant units while quantities such as x/m are purely numerical.

Table 1.2 The position coordinate x of the car in Figure 1.5 at various times t.

t/s	x/m
0	0
5	1.7
10	6.8
15	15
20	26
25	39
30	53
35	68
40	84
45	99
50	115
55	131
60	146

2.3 Position–time graphs

Tables do not give a very striking impression of how one thing varies with respect to another. A visual form of presentation, such as a graph, is usually much more effective. This is evident from Figure 1.6, which shows the graph obtained by plotting the data in Table 1.2 and then drawing a smooth curve through the resulting points.

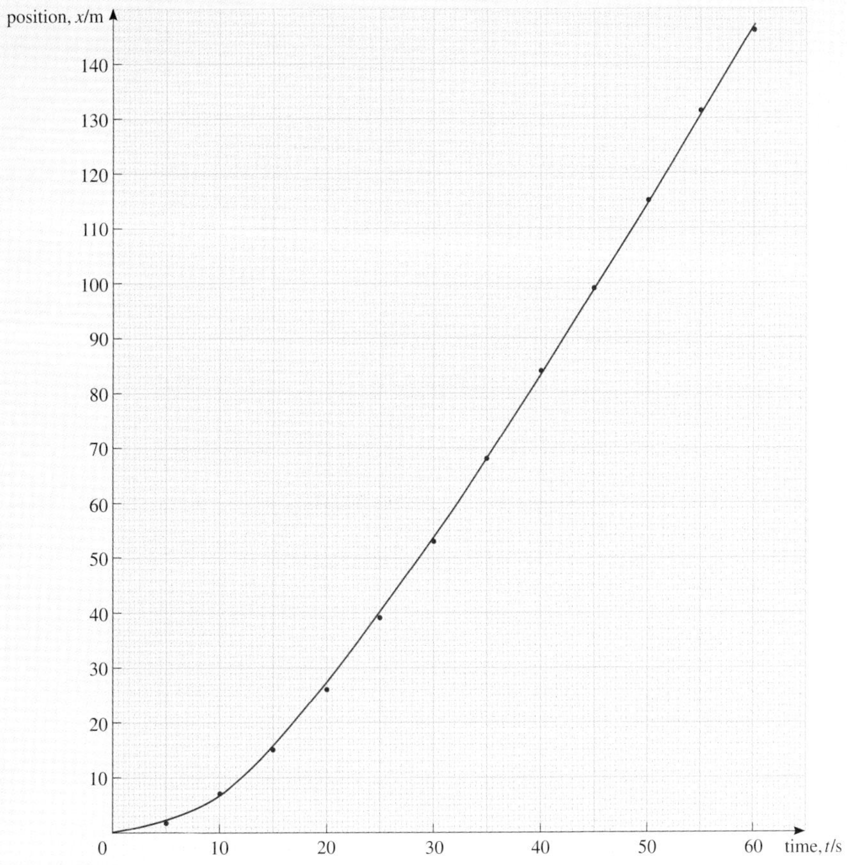

Figure 1.6 A position–time graph based on the results in Table 1.2.

The smooth curve drawn in Figure 1.6 is called the **position–time graph** of the car's motion. It can be used to read off the position of the car at any instant of time, or to find when the car passes a certain point. In graphs such as this, it is conventional to plot the time t along the horizontal axis. The vertical axis is used for the position x. This is just a standard way of displaying information about quantities that depend on time; it does *not* imply that the x-axis of Figure 1.5 is vertical!

Question 1.2 (a) Use Figure 1.6 to estimate the position of the car at $t = 32$ s. (b) Estimate the time at which the car reaches the position shown in Figure 1.5. ■

A position–time graph provides a very straightforward way of describing motion along a line. It is easy to construct from a table of measurements, and easy to use to determine details of the motion. However, you should realize that the appearance of the graph depends on the precise choice made for the x-axis. Figure 1.7 shows a new coordinate system in which the origin is 20 m to the right of the origin in Figure 1.5. As a result, the position coordinate of the car, measured in this new system, at any of the times listed in Table 1.2, will be 20 m *less* than the value given in the table, and the corresponding position–time graph will look like Figure 1.8. The car now starts at $x = -20$ m, and the steadily increasing x-coordinate only becomes positive after $t = 18$ s. Different choices of origin simply shift the position–time graph upwards or downwards, without changing its shape.

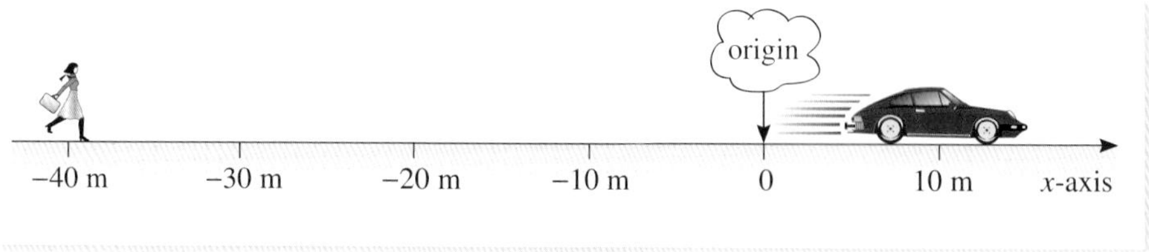

Figure 1.7 An alternative choice of x-axis with the origin moved 20 m to the right.

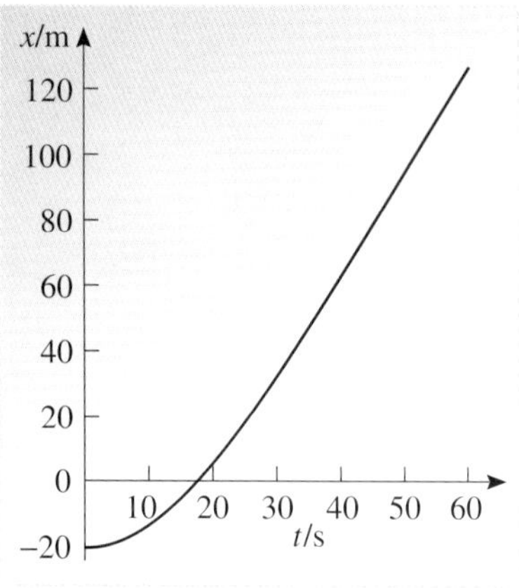

Figure 1.8 The position–time graph when the position of the car is measured relative to the x-axis of Figure 1.7.

Question 1.3 What change in the description of the motion would shift the position–time graph to the right or the left, without changing its shape? ■

2.4 Displacement–time graphs

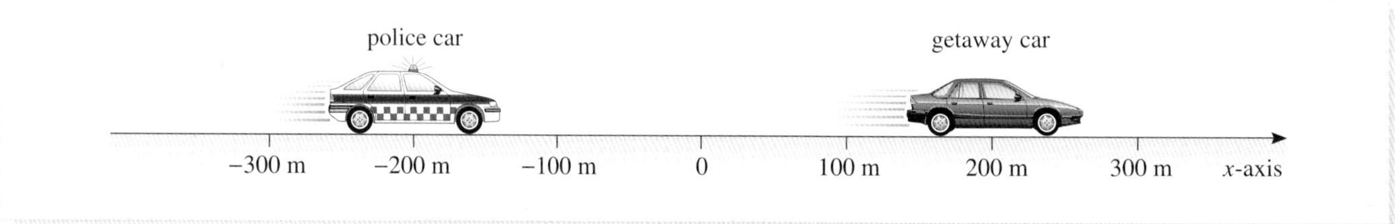

Figure 1.9 A police car in pursuit of a getaway car.

A particle's position, x, is always measured from the origin of the coordinate system. However, in describing real motions it is often important to know where something is located relative to a point other than the origin. Figure 1.9 shows a case in point; in a high-speed pursuit neither the police nor the robbers are likely to be very interested in their location relative to the origin, but both will be interested in the location of their own vehicle relative to the other. The physical quantity used to describe the location of one point relative to another is called **displacement**. In the case of Figure 1.9, the displacement of the getaway car from the police car is 400 m and the displacement of the police car from the getaway car is −400 m. In each case the displacement of a body is determined by subtracting the position coordinate of the reference body from the position coordinate of the body of interest. Thus the displacement (measured along the x-axis) of a particle with position x from a chosen reference point with position x_{ref} is given by

$$s_x = x - x_{ref}. \tag{1.1}$$

Notice that displacements, like positions, may be positive or negative depending on the direction of the displacement. Thus s_x is always positive when x is to the right of x_{ref} and is negative when x is to the left of x_{ref}. Also note that since displacements are measured along a definite axis it makes sense to represent them by a symbol, s_x, that includes a reference to that axis. This symbol may be read as 's subscript x' or, more simply, as 's sub x'.

A special case of Equation 1.1 is when the reference point x_{ref} is the origin. Then $x_{ref} = 0$ and $s_x = x$. Thus the displacement of a point from the origin is the position coordinate of the point. Once we know the displacement s_x of one object from another, it is easy to work out the **distance** s between those two objects. The distance is just the numerical value of the displacement, including the unit of measurement but ignoring any overall negative sign. So, when the displacement of the police car from the getaway car is $s_x = -400$ m, the distance between them is $s = 400$ m. We describe this simple relationship between distance and displacement by saying that the distance between two objects is given by the **magnitude** of the displacement of one from the other, and we indicate it mathematically by writing

$$s = |\,s_x\,|. \tag{1.2}$$

The two vertical bars | | constitute a **modulus sign** and indicate that you should work out the value of whatever they enclose and then take its magnitude (i.e. ignore any overall minus sign).

● In Figure 1.7, what is the displacement s_x of the car from the pedestrian, and what is the distance s between them?

❍ $s_x = 50$ m, and $s = 50$ m. ■

In many circumstances it is more valuable to plot a **displacement–time graph** rather than a position–time graph. In order to do this you either have to know the relevant displacements at various times, or you need to know enough about the positions of both the bodies involved to work out the displacements. Table 1.3 contains some plausible data about the positions of the two cars in Figure 1.9; use it to answer the following question.

Table 1.3 The position coordinates of the police car x_{pol} and getaway car x_{get} at various times t.

t/s	x_{pol}/m	x_{get}/m
0	−200	200
5	−115	237
10	−45	265
15	10	293
20	62	317
25	108	337
30	149	351
35	184	364
40	213	377
45	239	388
50	263	394
55	283	399
60	300	400

Question 1.4 Plot a graph to show how the displacement of the getaway car from the police car depends on time. ■

2.5 A note on graph drawing

There will be many occasions throughout your study of physics when you will need to draw graphs. This subsection gives some important guidelines for this activity.

1 *Decide* which is the **independent variable** and which the **dependent variable**. *Plot* the independent variable along the horizontal axis and the dependent variable along the vertical axis. This is purely a convention but is why, for instance, we usually plot the time along the horizontal axis of a position–time graph. It is the position that varies with time rather than the time that varies with position. Time is the independent variable since we can choose to make a measurement at any time. Position is the dependent variable.

Table 1.4 Positions for the pedestrian, x_{ped}, and the car, x_{car}, at times t.

t/s	x_{ped}/m	x_{car}/m
0	−20	30
10	−10	60
20	0	90
30	10	120
40	20	150
50	30	180
60	40	210

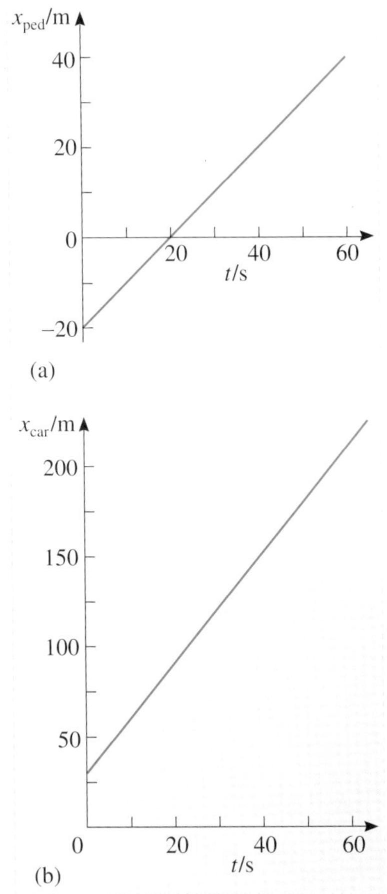

Figure 1.10 Position–time graphs for the positions of (a) a pedestrian and (b) a car, based on Table 1.4.

2 *Give* the graph a title e.g. distance versus time.

3 *Arrange* the axes so that the vertical axis increases in an upward direction and values along the horizontal axis increase to the right. This is simply a convention.

4 *Label* both axes to show which quantities are being plotted and include the units. By convention only pure numbers are plotted. The physical quantity must be divided by its units before being plotted. This means that each axis should be labelled as quantity/units. This is why in all the graphs we have drawn so far the axes have been labelled by time/s and position/m (or displacement/m).

5 *Fill* as much of the graph paper as reasonably possible. You will obtain greater accuracy if the graph is as big as possible. However, take care to use the graph paper sensibly. Graph paper usually has centimetre and millimetre squares, so it is straightforward to use 2 or 5 or 10 divisions on the paper to one physical unit. What you should avoid are multiples such as 3, 6, 7 ….

6 *Scale* the axes appropriately, especially if the numbers involved are either very large or very small. For example, if the values of time t range from 0 s to 1.0×10^{-5} s, then, rather than plotting t/s and inserting values such as 1.0×10^{-6}, 2.0×10^{-6}, etc. along the axis, it is usually more convenient to change the units to microseconds and plot $t/\mu s$; the values along the axis will then simply be 1, 2, 3, etc. It is also acceptable to label the axis $t/10^{-6}$ s rather than $t/\mu s$ if you prefer.

7 *Plot* the points clearly. If you use very small dots they may be confused with other marks on the paper. However, using very big dots is not a good idea since it is hard to tell the position of the centre. Some authors put the dots within small circles. In this course we simply use dots since it is easy to show them clearly in professionally drawn graphs.

8 *Draw* a straight line or smooth curve through the points plotted. The graphs that you draw will generally represent the smooth variation of one quantity with respect to another so a smooth curve is usually appropriate.

Question 1.5 How many of the above guidelines did you violate in answering Question 1.4? ■

3 Uniform motion along a line

3.1 Describing uniform motion

Uniform motion along a line is the very special kind of motion that occurs when an object moves with unvarying speed in a fixed direction. During a fixed period of time, such as one second, an object in uniform motion will always cover the same distance, no matter when the period begins. This is the kind of motion associated with traffic-free motoring along straight roads, with uninterrupted train journeys along straight tracks, and with unhindered straight and level flying (provided the distances involved are sufficiently small that the curvature of the Earth can be ignored).

Table 1.4 shows some typical values of position and time for a pedestrian and a car that are both in a state of uniform motion. The corresponding position–time graphs are shown in Figure 1.10a and b. As you can see, both are *straight-line graphs*, as is characteristic of uniform motion. The graphs indicate that in each case the quantities being plotted are related by an equation of the general form

$$x = At + B \qquad (1.3)$$

where A and B are constants. The two parts of Figure 1.10 simply correspond to different values of the constants A and B. In fact, as you will see later, by choosing appropriate values for A and B, Equation 1.3 can be used to represent any straight line that can be drawn on the position–time graph, except one that runs parallel to the vertical axis. In view of this you will not be surprised to learn that any equation of the general form $x = At + B$ may be referred to as the **equation of a straight line**, irrespective of the particular constants and variables that it involves.

Understanding the link between equations (such as Equation 1.3) and graphs (such as those in Figure 1.10) is of vital importance throughout physics. Broadly speaking, any equation that relates two variables can be represented as a graph, and any graph showing how one quantity varies with another can be represented by an equation. (There are exceptions to this broad statement, but they tend to be rather unphysical, and won't be considered here.) The skill of looking at a simple equation and visualizing it as a graph is one that is well worth developing. Many physicists find that visualization helps to bring equations to life and makes it possible to look 'into' equations rather than merely looking 'at' them. The equation of a straight line that describes uniform motion is a good place to start developing this skill.

In the next two subsections you will learn how the constants A and B in the equation of a straight line determine the slope and positioning of the corresponding line, and what those features represent physically in the case of a position–time graph.

Question 1.6 Use Table 1.5 to plot a graph showing how the displacement of the car from the pedestrian varies with time. Describe your graph in words and write down the general form of the equation that describes your graph. ■

Table 1.5 Displacement of the car from the pedestrian at time t, according to Table 1.4.

t/s	$(x_{car} - x_{ped})$/m
0	50
10	70
20	90
30	110
40	130
50	150
60	170

3.2 Constant velocity and the gradient of the position–time graph

Two things you will almost certainly want to know about any particle undergoing uniform motion are 'how fast is it travelling?' and 'in which direction is it moving?' The physical quantity that provides both these items of information is the particle's **velocity**. This is defined as the rate of change of the particle's position with respect to time, and has a constant value for each case of uniform motion along a line.

velocity = rate of change of position with respect to time.

In the case of the car whose position–time graph is shown in Figure 1.10b, if we choose two different times, say, $t_1 = 40$ s and $t_2 = 50$ s, then in the time interval between t_1 and t_2, the car moves from $x_1 = 150$ m to $x_2 = 180$ m. It follows that the rate of change of position of the car is given by the ratio

$$\frac{\text{change of position}}{\text{change of time}} = \frac{x_2 - x_1}{t_2 - t_1} = \frac{(180 - 150)\,\text{m}}{(50 - 40)\,\text{s}} = \frac{30}{10}\,\text{m}\,\text{s}^{-1} = 3.0\,\text{m}\,\text{s}^{-1}.$$

We therefore say that the velocity of the car is 3.0 metres per second along the x-axis, which we write using the abbreviation $\text{m}\,\text{s}^{-1}$.

More generally, if any particle moves uniformly along the x-axis, so that its position–time graph is a straight line, then the constant velocity v_x ('v sub x') of that particle is given by

$$v_x = \frac{x_2 - x_1}{t_2 - t_1}$$

where x_1 is the particle's position at time t_1, and x_2 is its position at time t_2. Note that the velocity may be positive or negative, depending on whether the particle's position coordinate is increasing or decreasing with time. Also note that for uniform motion the velocity is independent of the particular values of t_1 and t_2 that are chosen.

Question 1.7 Determine the velocity of the car by making measurements on the graph in Figure 1.10b in the interval from 50 s to 60 s. ■

In everyday speech the terms *velocity* and *speed* are used interchangeably. However, in physics, the term **speed** is reserved for the *magnitude* of the velocity, i.e. its value neglecting any overall minus sign. So, if a particle has velocity $v_x = -5\,\mathrm{m\,s^{-1}}$, then its speed is $v = 5\,\mathrm{m\,s^{-1}}$, where

$$v = |v_x|. \tag{1.4}$$

Speed is a *positive* quantity telling us how rapidly the particle is moving but revealing nothing about its direction of motion.

The subscript x in v_x may seem a bit cumbersome, but it is an essential part of the notation. The subscript reminds us that we are dealing with motion along the x-axis. In later work, we will need to use two or sometimes three axes, so by including the subscript at this stage, we will be able to use the same notation throughout. Be careful to include the subscript x in your written work and make sure that it is small enough and low enough to be read as a subscript — don't risk having v_x misinterpreted as vx, i.e. v times x.

Another piece of shorthand that you will find useful concerns the upper case Greek letter delta, Δ. If a quantity such as x changes its value from x_1 to x_2, then the *change* in the value of x is conventionally written as Δx, and read as 'delta ex'. Thus

$$\Delta x = x_2 - x_1.$$

So, Δx and Δt mean the changes in x and t; they do *not* mean Δ times x or Δ times t. You should always think of the Δ and the symbol that follows it as a single entity; the Δ symbol by itself has no quantitative meaning. Using this notation, the expression for v_x above can be written as

$$v_x = \frac{\Delta x}{\Delta t} = \frac{x_2 - x_1}{t_2 - t_1}. \tag{1.5}$$

In graphical terms, the velocity $\Delta x/\Delta t$ of a uniformly moving particle is indicated by the *slope* of its position–time graph. The steepness of the line represents the speed of the particle, while the orientation of the line — bottom left to top right, or top left to bottom right — indicates the direction of motion. If the graph of x against t is a straight line, then the ratio of the change in x to the corresponding change in t (that is $\Delta x/\Delta t$) is called the **gradient** of the graph. A line that slopes from bottom left to top right indicates that a positive change in t corresponds to a positive change in x, consequently such a line has positive gradient. Similarly, a line sloping from top left to bottom right indicates that a positive change in t corresponds to a negative change in x and consequently to a negative gradient. This gives us another way of describing velocity:

the velocity of a particle = the gradient of its position–time graph.

Question 1.8 Figure 1.11 shows the position–time graphs for four different objects (A, B, C and D) each moving uniformly with a different constant velocity. The position and time scales are the same in each case. (a) List the objects in order of increasing speed. (b) Which of the objects have positive velocity? (c) List the objects in order of increasing velocity. ■

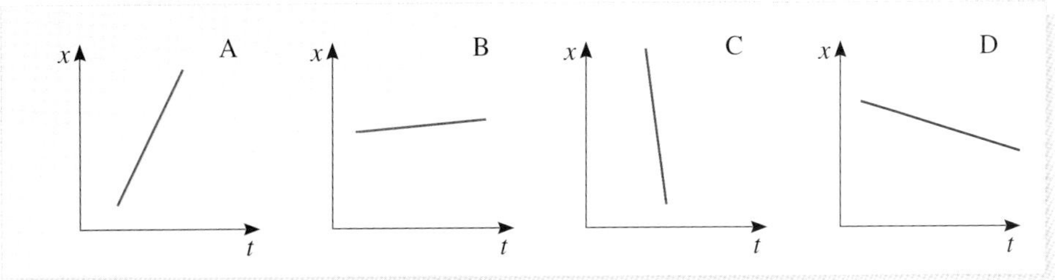

Figure 1.11 Straight line position–time graphs for Question 1.8.

The gradient of a graph is one of the most important concepts of this chapter. It is important that you should be able to evaluate gradients, and that you should be able to distinguish graphs with positive gradients from those with negative gradients. It is also important that you should be able to interpret gradients physically as rates of change. In this discussion of position–time graphs the gradient represents velocity, but in other contexts, where quantities other than x and t are plotted, the gradient may have a very different interpretation. Figure 1.12, for example, shows how the average temperature T of the atmosphere depends on the height h above sea-level. The gradient of this graph, $\Delta T/\Delta h$, describes the rate at which temperature changes with height.

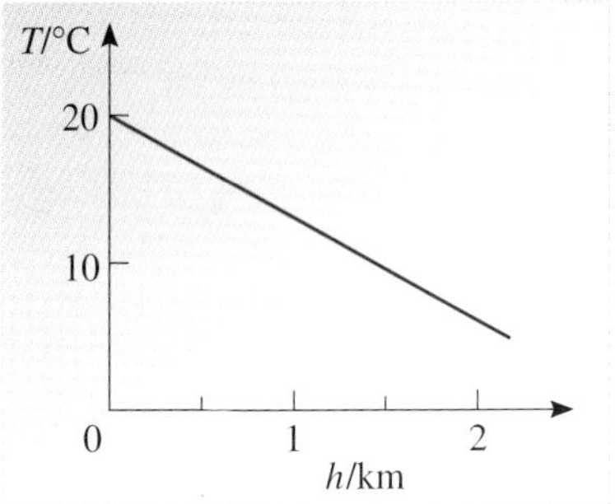

Figure 1.12 A graph of average atmospheric temperature T against height h above sea-level.

Question 1.9 Estimate the gradient of the graph in Figure 1.12. (The use of the word 'estimate' implies that you do not need to take great care over the values you read from the graph, but you should take care over matters such as signs and units of measurement.) ■

3.3 Initial position and the intercept of the position–time graph

The uniform motion of a particle is such a simple form of motion that apart from enquiring about the particle's velocity, the only other kinematic question you can ask is 'where was the particle at some particular time?' The most common way of answering this question is to specify the **initial position** of the particle, that is, its position at time $t = 0$ s.

Although it is common to refer to the position at $t = 0$ as the 'initial position' it is also possible, and sometimes more convenient, to associate the initial position with some other time.

The initial position of a uniformly moving particle is easily determined from its position–time graph. It's just the value of x when $t = 0$, i.e. the value of x at which the straight line crosses the vertical axis through the origin. In Figure 1.10a, for example, it is $x = -20$ m. This value is generally referred to as the **intercept**.

One point to bear in mind though; it is sometimes advantageous to draw position–time graphs that do not include the origin (for instance, you might be asked to draw a graph for the period from $t = 100$ s to $t = 110$ s). When dealing with such graphs do not make the mistake of thinking that the value at which the line crosses the vertical axis is the intercept. You can only read the intercept directly from the graph if the vertical axis passes through the zero value on the horizontal axis.

A little thought should convince you that the gradient and the intercept of a straight line entirely determine that line. In the same way, the velocity and the initial position of a uniformly moving particle entirely determine the motion of that particle. In the next subsection you will learn how these graphical and physical statements can be represented algebraically, in terms of equations.

3.4 The equations of uniform motion

It has already been said that the straight-line graph of any uniform motion can be represented by an equation of the general form

$$x = At + B \qquad \text{(Eqn 1.3)}$$

where A and B are constants. Different cases of uniform motion simply correspond to different values for the constants A and B. Let us now investigate this equation to see how it conveys information about gradients and intercepts, or equivalently, uniform velocities and initial positions.

To start with, note that according to Equation 1.3, the position of the particle when $t = 0$ is just $x = B$. Thus B represents the initial position of the particle, the intercept of its position–time graph.

In a similar way, note that according to Equation 1.3, at time $t = t_1$ the position of the particle, let's call it x_1, is given by $x_1 = At_1 + B$; and at some later time $t = t_2$ the position of the particle is $x_2 = At_2 + B$. Now, as we have already seen, the velocity of a uniformly moving particle is defined by

$$v_x = \frac{\Delta x}{\Delta t} = \frac{x_2 - x_1}{t_2 - t_1} \qquad \text{(Eqn 1.5)}$$

so, substituting the expressions for x_1 and x_2 that we have just obtained from Equation 1.3 we find that in this particular case

$$v_x = \frac{x_2 - x_1}{t_2 - t_1} = \frac{(At_2 + B) - (At_1 + B)}{t_2 - t_1} = \frac{A(t_2 - t_1)}{t_2 - t_1} = A.$$

Thus, the constant A in Equation 1.3 represents the particle's velocity, i.e. the gradient of its position–time graph.

We can now read the equation of a straight line, interpreting it graphically or physically. It is

	gradient	intercept	graphical interpretation
	$\Uparrow$	$\Uparrow$	
equation of a straight line	$x = At$	$+ \; B$	
	$\Downarrow$	$\Downarrow$	
	velocity	initial position	physical interpretation

The physical significance of the constants A and B can be emphasized by using the symbols v_x and x_0 in their place. Doing this, we obtain the standard form of the **uniform motion equations**

$$x = v_x t + x_0 \qquad \text{(1.6a)}$$

$$v_x = \text{constant}. \qquad \text{(1.6b)}$$

So, the uniform motion of the car we have been considering, which has velocity $v_x = 30\,\text{m}\,\text{s}^{-1}$ and initial position $x_0 = 30\,\text{m}$, can be described by the equation

$$x = (30\,\text{m}\,\text{s}^{-1})t + (30\,\text{m}).$$

This contains just as much information as the graph in Figure 1.10b.

Question 1.10 Write down the equation that describes the uniform motion of the pedestrian who was the subject of Figure 1.10a. ■

A pictorial interpretation of Equation 1.6a is given in Figure 1.13.

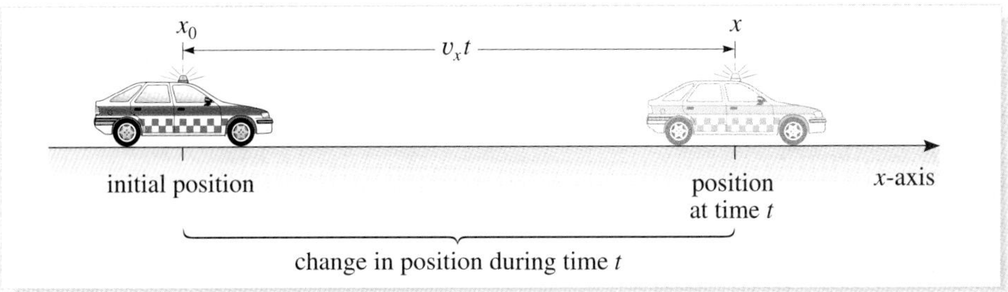

Figure 1.13 A pictorial interpretation of the equation of uniform motion $x = v_x t + x_0$.

Equations have the great advantage that they can be rearranged to make them fit the problem at hand. For instance, if you were flying at constant velocity from Paris to Rome, you might well be more interested in your displacement from Paris rather than your position in some arbitrary coordinate system. In mathematical terms, you might be more interested in $s_x = x - x_0$ rather than x itself. Equation 1.6a can be rearranged to suit this interest by subtracting x_0 from each side:

$$x - x_0 = (v_x t + x_0) - x_0$$

and this may be rewritten in the compact form

$$s_x = v_x t. \tag{1.7}$$

Remembering that v_x is a constant when the motion is uniform, Equation 1.7 is equivalent to Equation 1.6a. It relates the displacement from the initial position directly to the time of flight and thus facilitates working out the distance travelled,

$$s = |s_x| = |v_x t|. \tag{1.8}$$

The crucial point to remember when manipulating an equation is that *both sides of an equation represent the same value*. So, if you add or subtract a term on one side of an equation, you must add or subtract an identical term on the other side. The same principle applies to multiplication and division, and to other operations such as taking magnitudes, squaring or taking square roots. (In the latter case you must remember that a positive quantity generally has two square roots; 2 is a square root of 4, but so is −2.) You will learn more about rearranging equations as you read this book, but there are two rules to bear in mind throughout:

1 When one side of an equation consists of several terms added together, each of those terms must be treated in the same way. (It's no good multiplying the first term on one side of an equation by some quantity if you forget to do the same to all the other terms on that side.)

2 When dividing both sides of an equation by some quantity you must ensure that the quantity is *not* zero. (For instance, it is only legitimate to divide both sides of Equation 1.7 by $x - x_0$ *if* $x \neq x_0$, i.e. if x is *not equal* to x_0.)

The following question is quite straightforward, and there are many ways of answering it, but each will require you to manipulate the uniform motion equations to some extent. As you work out your answer think carefully about each of the procedures you are performing and write it out in words.

Question 1.11 When the time on a certain stop-watch is 100 s, a vehicle is positioned at $x = -2$ m with respect to a certain one-dimensional coordinate system. If the velocity of the vehicle is -12 m s^{-1}, find its position when the stop-watch shows 250 s. ■

3.5 Velocity–time and speed–time graphs

Just as we may plot the position–time graph or the displacement–time graph of a particular motion, so we may plot a **velocity–time graph** for that motion. By convention, velocity is plotted on the vertical axis (since velocity is the dependent variable) and time (the independent variable) is plotted on the horizontal axis. In the special case of uniform motion, the velocity–time graph takes a particularly simple form — it is just a horizontal line, i.e. the gradient is zero. Examples are given in Figure 1.14; notice that the velocity can be positive or negative, depending on the direction of motion.

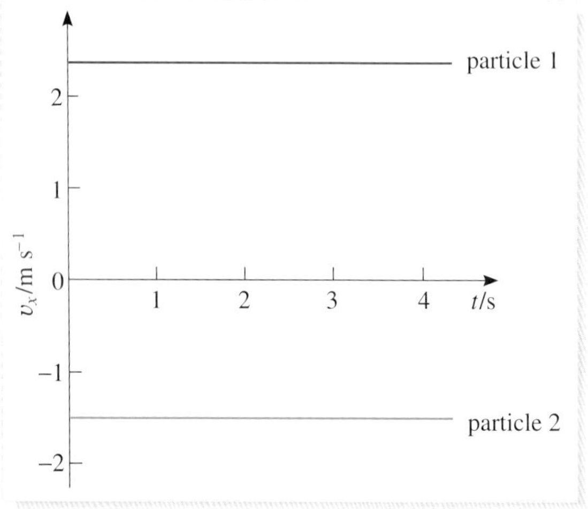

Figure 1.14 Velocity–time graphs for uniform motion.

Rather than plotting the velocity it is sometimes useful to plot the magnitude of the velocity, in other words, the speed. The resulting plots are **speed–time graphs** and examples are shown in Figure 1.15. Notice how all of the speeds are positive; in particular, the velocity of -1.5 m s^{-1} in Figure 1.14 corresponds to a speed of 1.5 m s^{-1} in Figure 1.15.

Clearly, a speed–time graph provides less information than a velocity–time graph, but it may be sufficient. The next time you buy a Ferrari, you may well enquire about its top speed, but you are unlikely to ask 'in which direction?'

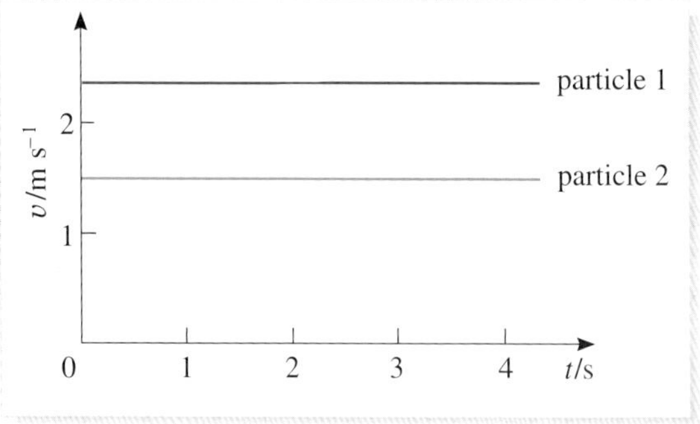

Figure 1.15 Speed–time graphs for uniform motion, corresponding to the velocity–time graphs shown in Figure 1.14.

3.6 The signed area under a constant velocity–time graph

There is a simple feature of uniform velocity–time graphs that will be particularly useful to know about when we come to consider non-uniform motion in the next section. It concerns the relationship between the velocity–time graph and the change in position over a given time interval. Consider the following problem. A vehicle travels at a velocity $v_x = 12\,\mathrm{m\,s^{-1}}$ for 4 s. By how much does its position change over that interval?

The answer, from Equation 1.7, is 48 m. However, for our present purposes it is more instructive to work from the definition of uniform velocity (Equation 1.5), which may be rearranged by multiplying both sides of the equation by $(t_2 - t_1)$ to give

$$x_2 - x_1 = v_x(t_2 - t_1).$$

This tells us that the change in position during a given time interval is equal to the velocity multiplied by the time interval. So, a vehicle which travels at a constant velocity $v_x = 12\,\mathrm{m\,s^{-1}}$ over a time interval $\Delta t = 4$ s will change its position coordinate by $\Delta x = 48$ m.

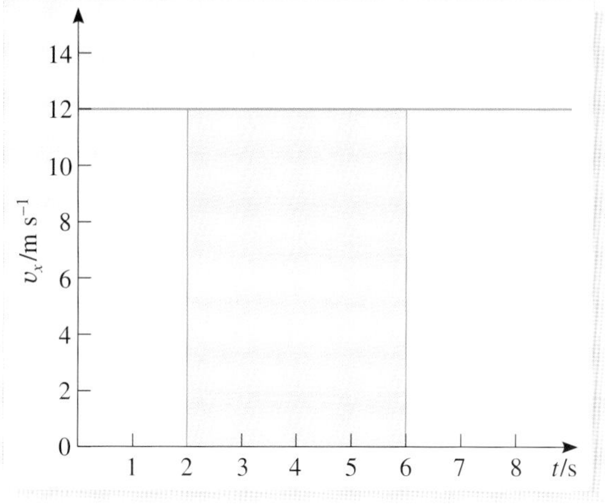

Figure 1.16 The area under a velocity–time graph.

Now look at the velocity–time graph for this vehicle, which is given in Figure 1.16. The colour-shaded region corresponding to the 4 s interval between $t = 2$ s and $t = 6$ s is an example of an *area under the graph*: that is an area bounded by the plotted line, the horizontal axis and two vertical lines. The important point to note about this region is its area — not its physical area in terms of $\mathrm{m^2}$ of paper, but rather its area in terms of the units used on the axes of the graph. The area of a rectangle is the product of the lengths of two adjacent sides. So in Figure 1.16 the area under the graph is

$$v_x \times \Delta t = (12\,\mathrm{m\,s^{-1}}) \times (4\,\mathrm{s}) = 48\,\mathrm{m}.$$

Clearly, the area under the velocity–time graph, between the specified times, is exactly equal to the change in position coordinate between those times. In case you are worried that this relationship will break down if the velocity is negative, as in Figure 1.17, it should be added that areas that hang below the horizontal axis are conventionally regarded as negative areas. In this sense the area under a graph is a signed quantity that may be positive or negative; indeed, it is often referred to as the **signed area under a graph**.

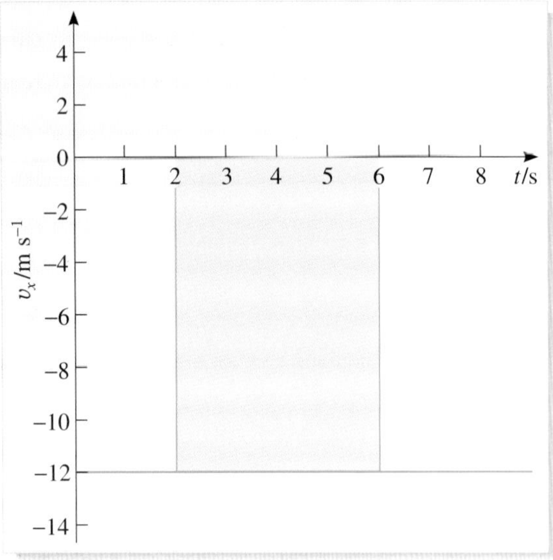

Figure 1.17 The area under a velocity–time graph is a signed quantity that may be negative.

Recognizing the equality between the signed area under a uniform velocity–time graph and a change in position coordinate is not particularly helpful in itself, but it does pave the way for further developments, as does the next question.

Here we are using a double subscript on the symbol v. The subscript x denotes the x-direction as usual, and the 1 and 2 refer to the different stages of the motion.

Question 1.12 An athlete walks with a constant velocity $v_{x1} = 1\,\text{m s}^{-1}$ for 20 s, and then abruptly starts running with a constant velocity $v_{x2} = 10\,\text{m s}^{-1}$, a velocity that is maintained for a further 20 s. What is the area under the corresponding velocity–time graph, and what physical interpretation could you give to that area? ■

3.7 A note on straight-line graphs and their gradients

We end this section by reviewing some of the important features of straight-line graphs, though we do so in terms of two general variables z and y, rather than x and t, in order to emphasize their generality. If the graph of z against y is a straight line of the kind shown in Figure 1.18, then z and y are related by an equation of the form

$$z = my + c \tag{1.9}$$

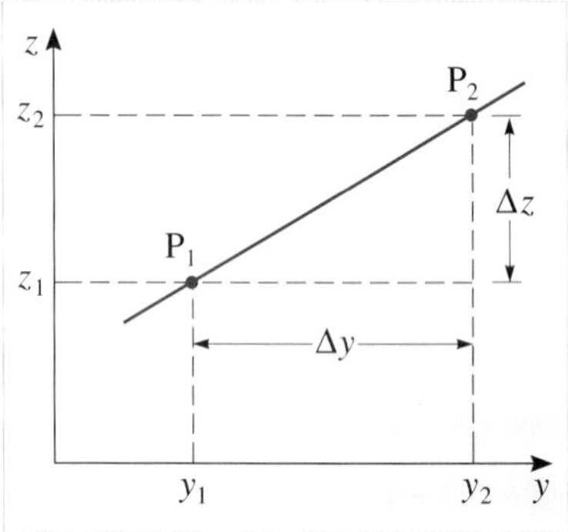

Figure 1.18 A straight-line graph of z against y.

where m and c are constants. Here c represents the intercept of the graph and is equal to the value of z at which the plotted line crosses the z-axis (provided the z-axis passes through $y = 0$). The constant m represents the gradient of the graph and is obtained by dividing the change in z by the corresponding change in y, taking full account of the sign of each change.

You may find it useful to remember that the gradient of a graph is given by its 'rise' over its 'run'.

Thus $$\text{gradient} = \frac{\Delta z}{\Delta y} = \frac{z_2 - z_1}{y_2 - y_1}. \tag{1.10}$$

There are several points to notice about this definition.

1. It applies only to straight-line graphs.
2. The units of the gradient are the units of z divided by the units of y.
3. The gradient of a straight-line graph is a constant (a number, multiplied by an appropriate unit). The same constant is obtained, no matter which two points (see P_1 and P_2 in Figure 1.18) are used to determine it.
4. The gradient of a straight-line graph can be positive, negative or zero. Equation 1.10 assigns a positive gradient to a graph sloping from bottom left to top right, as in Figure 1.18 for example, and a negative gradient to a graph sloping from top left to bottom right, as in Figure 1.12. A horizontal line has zero gradient.

Although the appearance of a graph can be changed by plotting the points on different scales (compare Figures 1.19a and b for example) the gradient, defined by Equation 1.10, is independent of the shape or size of the graph paper or display screen.

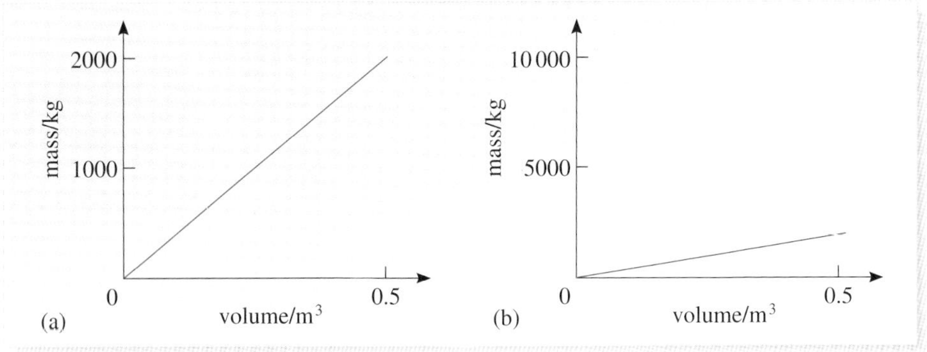

Figure 1.19 Two graphs constructed from the same data but plotted with different scales. The gradients are the same, 2.5×10^{-4} kg m^{-3}.

4 Non-uniform motion along a line

4.1 Instantaneous velocity

Uniform motion is simple to describe, but is rarely achieved in practice. Most objects do not move at a precisely constant velocity. If you drop an apple it will fall downwards, but it will pick up speed as it does so (Figure 1.20), and if you drive along a straight road you are likely to encounter some traffic that will force you to vary your speed from time to time. For the most part, real motions are **non-uniform motions**.

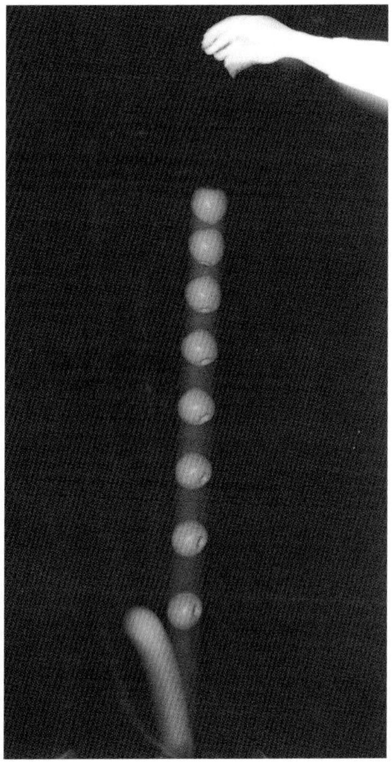

Figure 1.20 A falling apple provides an example of non-uniform motion. A sequence of pictures taken at equal intervals of time reveals the increasing speed of the apple as it falls.

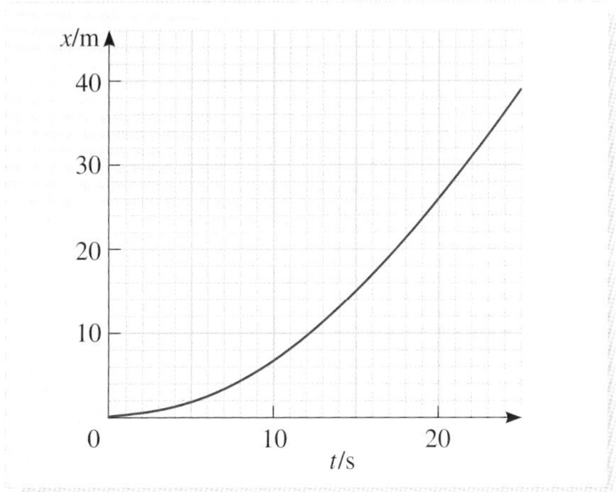

Figure 1.21 The position–time graph for a car accelerating from rest.

Figure 1.21 shows the position–time graph of an object that has an increasing velocity over the period $t = 0$ to $t = 20$ s; a car accelerating from rest. As you can see, the position–time graph is curved. There is relatively little change in position during the first few seconds of the motion but as the velocity increases the car is able to change its position by increasingly large amounts over a given interval of time. This is shown by the increasing steepness of the graph. In everyday language we would say that the graph has an increasing gradient, but you saw in the last section that the term gradient has a precise technical meaning in the context of straight-line graphs. Is it legitimate to extend this terminology to cover curved graphs, and if so, how exactly should it be done?

Extending the concept of gradient to the case of curved graphs is actually quite straightforward. The crucial point to recognize is that if you look closely enough at a small part of a smooth curve, then it generally becomes indistinguishable from a straight line. (In a similar way, the surface of the Earth is clearly curved when viewed from space, but each region is approximately flat when seen close-up.) So, if we choose a point on a curve we can usually draw a straight line passing through that point which has the same slope as the curve at the point of contact. This straight line is said to be the **tangent** to the curve at the point in question. Now, we already know how to determine the gradient of a straight line, so we can define the gradient at any point on a curve to be the gradient of the tangent to the curve at that point, provided the curve is sufficiently smooth that a tangent exists.

Figure 1.22 repeats the position–time graph of the accelerating car, but this time tangents have been added at $t = 5$ s and $t = 10$ s. The gradient of any such tangent represents a velocity and is referred to as the **instantaneous velocity** of the car at the relevant time. At least, that's what it should be called; in practice the word 'instantaneous' is often omitted, so references to 'velocity' should generally be taken to mean 'instantaneous velocity'. Allowing ourselves this informality, we can say:

velocity at time t = gradient of position–time graph at time t.

Although we have had to extend the meaning of gradient, we can still regard it as a measure of the rate of change of one variable with respect to another, so we can also say:

velocity at time t = rate of change of position with respect to time at time t.

Question 1.13 Estimate the (instantaneous) velocity of the car at $t = 5$ s and at $t = 10$ s (from Figure 1.22), and write down your answers taking care to distinguish one velocity from the other. ■

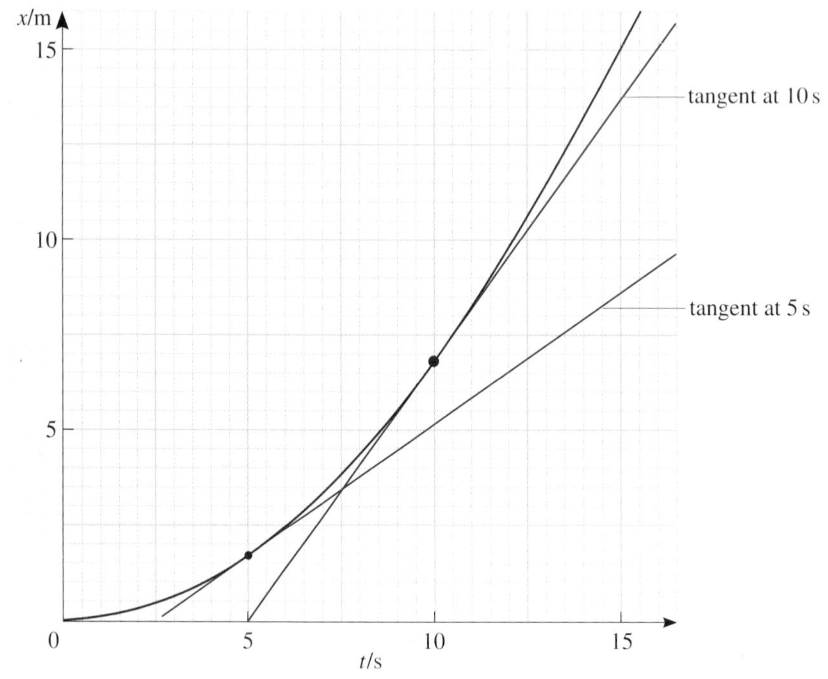

Figure 1.22 The instantaneous velocity at $t = 5$ s and at $t = 10$ s is determined by the gradient of the tangent to the position–time graph at each of those times.

How did you distinguish the velocity at $t = 5$ s from the velocity at $t = 10$ s? The conventional method is to use a common symbol for velocity, v_x, but to follow it by the relevant value of time enclosed in parentheses, as in $v_x(5\,\text{s})$ and $v_x(10\,\text{s})$. This notation can also be used to indicate the velocity at any time t, by writing $v_x(t)$, even if the value of t is unspecified.

It is important to remember that $v_x(t)$ represents the (instantaneous) velocity at time t. It does not mean v_x multiplied by t.

You will not be surprised to learn that the positive quantity $|v_x(t)|$ representing the magnitude of the instantaneous velocity at time t is called the **instantaneous speed** at time t. If we denote this by $v(t)$, we can write

$$v(t) = |v_x(t)|. \qquad (1.11)$$

Speed and velocity are measured in the same units, m s^{-1}. Some typical values of physically interesting speeds are listed in Figure 1.23.

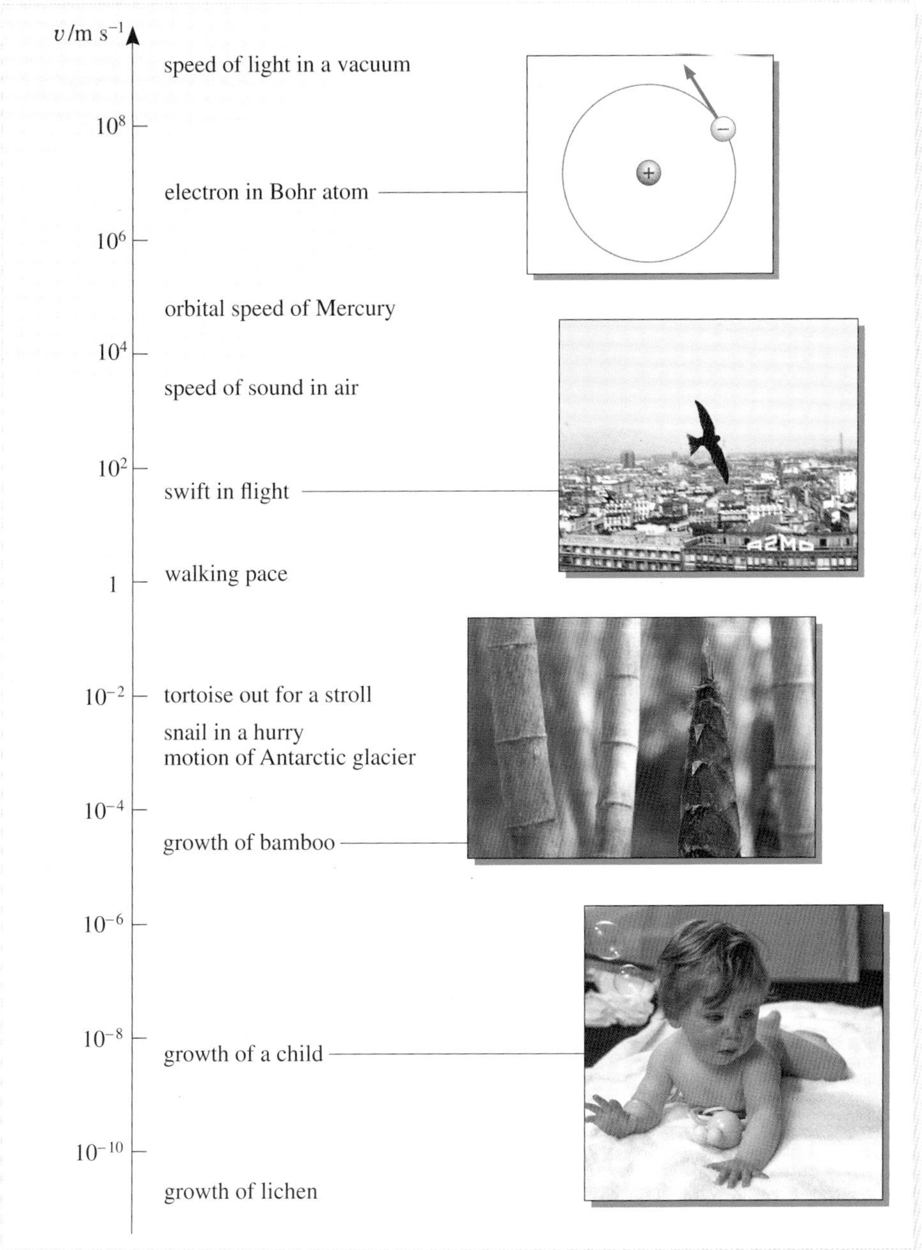

Figure 1.23 Some physically interesting speeds.

4.2 Instantaneous acceleration

The procedure of Question 1.13 for determining the instantaneous velocity of the car can be carried out for a whole set of different times and the resulting values of v_x can be plotted against t to form a graph. This has been done in Figure 1.24, which shows how the velocity varies with time. At time $t = 0$ s, the car has zero velocity because it starts from rest. At later times, the velocity is positive because the car moves in the direction of increasing x. The velocity increases rapidly at first, as the car picks up speed. Subsequently, the velocity increases more slowly, and eventually the car settles down to a steady velocity of just over 3 m s^{-1}. We have already come across graphs of this general type in Section 3; they are known as velocity–time graphs. The crucial new feature here is that the velocity now depends on time.

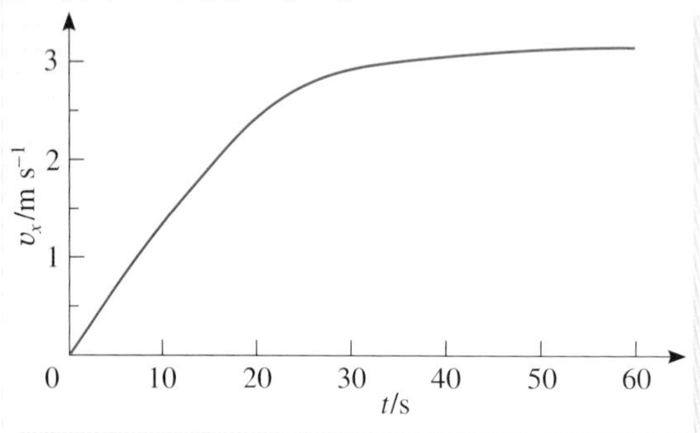

Figure 1.24 A velocity–time graph for the moving car.

The time dependence of velocity can have dramatic consequences. If you are on board a train, moving at a constant velocity, you might not even be aware of your motion and you will have no difficulty in, say, drinking a cup of coffee. However, drinking coffee can become distinctly hazardous if the driver suddenly changes the velocity of the train by putting on the brakes. In such situations the important physical quantity is the *rate of change of velocity with respect to time*, as measured by the gradient of the velocity–time graph. This is the quantity that we usually call acceleration, though once again it should more properly be called **instantaneous acceleration**. Thus

acceleration at time t = rate of change of velocity with respect to time at time t

or, if you prefer

acceleration at time t = gradient of velocity–time graph at time t.

Acceleration is a key idea in physics. It was Newton's recognition of the crucial role that acceleration played in determining the link between motion and force that formed the centrepiece of the Newtonian revolution. The detailed study of that revolution would take us too far from our present theme. For the moment let's concentrate on some basic questions about acceleration itself. In particular, in what units should acceleration be measured, and what are typical values of acceleration in various physical contexts? The first of these you can answer for yourself.

● What are suitable SI units for the measurement of acceleration?

❍ Since acceleration is the rate of change of velocity with respect to time, the units of acceleration are the units of velocity (m s^{-1}) divided by the units of time (s), so acceleration is measured in metres per second per second, which is abbreviated to m s^{-2}. ■

As for typical values of acceleration, some physically interesting values are shown in Figure 1.25.

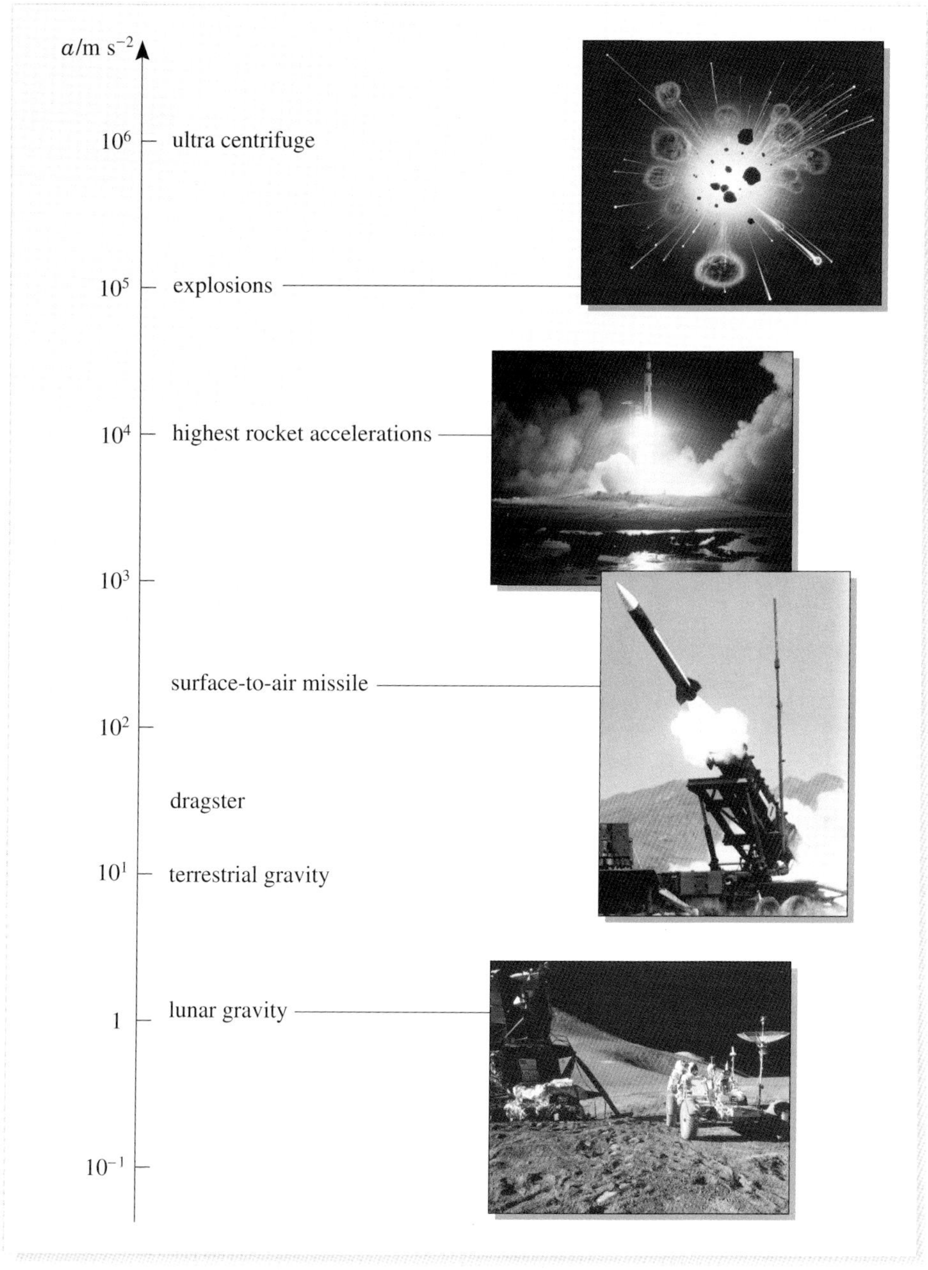

Figure 1.25 Some physically interesting values of acceleration.

When dealing with non-uniform motion along the x-axis, the symbol $a_x(t)$ is normally used to denote instantaneous acceleration. As usual, $a_x(t)$ will be positive if the velocity is increasing with time, though, as you will see below, this statement needs careful interpretation. In contrast to the relationship between velocity and speed, there is no special name for the magnitude of an acceleration, though we shall use the symbol $a(t)$ for this quantity, so we may write

$$a(t) = |a_x(t)|. \quad (1.12)$$

In physics, the concept of acceleration is precise and quantitative. It is important to realize that this precise definition differs, in some respects, from everyday usage. In ordinary speech, 'accelerating' is a synonym for 'speeding up'. This is not true in physics. In physics, a particle accelerates if it changes its velocity in *any* way. A particle travelling along a straight line may accelerate by speeding up *or* by slowing down. It is tempting to suppose that a positive acceleration corresponds to speeding up and a negative acceleration corresponds to slowing down, but this is not always true either, as the following exercise shows.

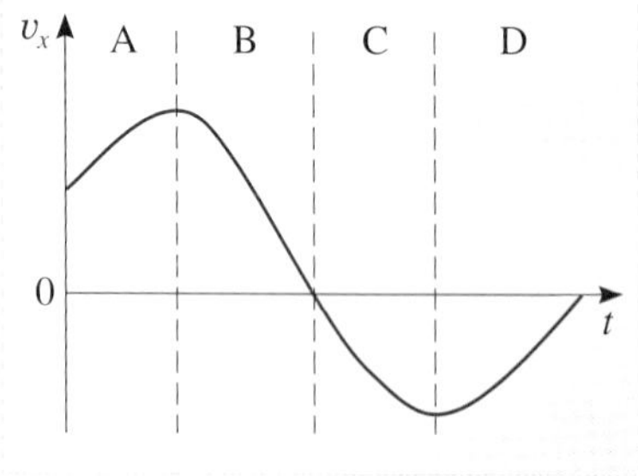

Figure 1.26 The velocity–time graph used in Question 1.14.

Question 1.14 The velocity–time graph of Figure 1.26 is divided into four regions, marked A–D. (a) In which regions does the particle move in the direction of increasing x? In which regions is it moving in the direction of decreasing x? (b) In which regions does the particle speed up? In which regions does it slow down? (c) In which regions does the particle have positive acceleration? In which regions does the particle have negative acceleration? ■

So, for a particle with negative velocity, an increase in velocity (a positive acceleration) may result in a decrease in speed. You may think that it is a nuisance for physics to use words in such a non-standard way. However, the definitions of velocity and acceleration given in this chapter are essential if we are to develop a study of kinematics that is both simple and comprehensive. It does mean, however, that you must be careful in using words like velocity, speed and acceleration. In everyday speech, the word deceleration is used to mean 'slowing down', but this term is seldom encountered in physics since it is already covered by the scientific definition of acceleration.

4.3 A note on functions and derivatives

This subsection introduces two crucially important mathematical ideas, *functions* and *derivatives*, both of which are used throughout physics.

> Open University students should leave the text at this point and use the multimedia package *Functions and derivatives*. When you have completed this activity you should return to this text. The activity will occupy about one hour.

Functions and the function notation

In Figure 1.21, the position x of the car depends on the time t. The graph associates a particular value of x with each value of t over the plotted range. In other circumstances we might know an equation that associates a value of x with each value of t, as in the case of the equation $x = At + B$ that we discussed in Section 3. You can invent countless other ways in which x depends on t: for instance $x = At^2 + B$ or $x = At + Bt^2$. In all such cases we describe the dependence of x on t by saying that x is a **function** of t. This terminology is widely used and is certainly not restricted to x and t. In the example of non-uniform motion we have just been discussing, the instantaneous velocity v_x is a function of time and so is the instantaneous acceleration a_x.

Generally, if the value of a quantity f is determined by the value of another quantity y, then we say that f is a function of y and we use the special notation $f(y)$ to emphasize this relationship. Although we have only just defined what we mean by a function we have already been using this notation, as in $v_x(t)$ and $a_x(t)$, for some time.

This function notation has two great merits:

1 Writing $f(y)$, provides a clear visual reminder that f depends on y in a well-defined way. If we happen to know the equation that relates f to y, say $f = y^2$, then we can show this explicitly by writing $f(y) = y^2$.

2 If we want to indicate the value of f that corresponds to a particular value of y, it is easy to do so. For example, the value of $f(y)$ at $y = 2$ can be written $f(2)$. We call $f(2)$ the *function value* at $y = 2$. Of course, in order to be able to write $f(2)$ as a number we would have to know the explicit form of $f(y)$. For example, if $f(y) = y^2$, then we can say $f(2) = 2^2 = 2 \times 2 = 4$.

The only serious disadvantage of the function notation is that you may confuse $f(y)$ with $f \times y$. Be careful! If f is a function, then $f(y)$ means f is a function of y; it does *not* mean $f \times y$.

Some functions arise repeatedly and are given special names so that they can be easily identified. You will be familiar with some of these names, even if you are not yet fully familiar with the functions they describe. For instance, if you look at a scientific calculator (see Figure 1.27 for example) you will sometimes find that there are keys labelled sin, cos, log and e^x; each of these is the name of an important function that you will meet later. The corresponding calculator keys are actually called 'function keys'. Electronic calculators are constructed in such a way that when you key in a number and press a function key, the calculator works out the corresponding function value and displays it. If the value you keyed in is not within the acceptable range of input values for the function you selected (usually called the *domain* of the function), then the calculator will probably display an error message such as 'err'.

Figure 1.27 A calculator has in-built routines for evaluating basic functions such as x^2, $\sin(x)$, $\cos(x)$, $\log(x)$, etc. These are activated by pressing the function keys.

One particularly simple class of functions consists of functions of the form

$$f(y) = Ay^n \qquad (1.13)$$

where A is a constant and n is a positive whole number, such as 0, 1, 2, 3, ... Functions of this kind include **squares** $f(y) = y^2$, and **cubes** $f(y) = y^3$, which correspond to $A = 1$ with $n = 2$ and $n = 3$, respectively. The function that arises when $n = 0$ is especially noteworthy since, by convention, $y^0 = 1$, so $f(y) = A$ in this case, a constant. Thus, even a simple constant is just a special kind of function. In the case of a constant function each value of y is associated with the same value of $f(y)$, as in the case of the velocity–time graph of a uniformly moving object.

Next in complexity are sums of squares, cubes, etc. These functions are called **polynomial functions** and include the following special cases (where A, B, C and D are constants):

linear functions of the form $f(y) = Ay + B$
quadratic functions of the form $f(y) = Ay^2 + By + C$
cubic functions of the form $f(y) = Ay^3 + By^2 + Cy + D$.

In the case of uniform motion (described in Section 3), the position–time graph is a straight line described by an equation of the form $x = At + B$. It should now be clear that another way of describing this relationship is to say that the position is a *linear function* of the time, $x(t) = At + B$. In a similar way, the non-uniform motion of a test vehicle in the NASA drop-shaft described at the beginning of this chapter can be described by a *quadratic function* $x(t) = At^2 + Bt + C$, and many other forms of one-dimensional motion can be described, at least approximately, by suitably chosen polynomial functions.

Derived functions and derivative notation

Given the function $x(t)$ that describes some particular motion, you could plot the corresponding position–time graph, measure its gradient at a variety of times to find the instantaneous velocity at those times and then plot the velocity–time graph. If you had some time left, you might go on to measure the gradient of the velocity–time graph at various times, and then plot the acceleration–time graph for the motion. This would effectively complete the description of the motion, but it would be enormously time consuming and, given the difficulty of reading graphs, not particularly accurate.

Fortunately this graphical procedure can usually be entirely avoided. Starting again from the function $x(t)$, there exists a mathematical procedure, called **differentiation**, that makes it possible to determine the velocity $v_x(t)$ directly, by algebra alone. We shall not try to describe the principles that underpin differentiation, but we will introduce the notation of the subject and list some of the basic results. To make this introduction as general as possible we shall initially consider a general function $f(y)$ rather than the position function $x(t)$.

The central idea is this:

Remember, the gradient of a graph at a given point is defined by the gradient of its tangent at that point.

> Given a function $f(y)$ it is often possible to determine a related function of y, called the **derived function**, with the property that, at each value of y, the derived function is equal to the gradient of the graph of f against y at that same value of y.

The derived function is usually referred to as the **derivative** of f with respect to y (often abbreviated to *derivative*) and may be represented by the symbol $\dfrac{\mathrm{d}f}{\mathrm{d}y}$ or, more formally $\dfrac{\mathrm{d}f(y)}{\mathrm{d}y}$. The $\dfrac{\mathrm{d}f}{\mathrm{d}y}$ notation is reminiscent of the $\dfrac{\Delta x}{\Delta t}$ notation that was used when discussing the gradient of a straight line and thus provides a clear reminder of the link between the derived function and the gradient of the f against y graph. However, it is important to remember that $\dfrac{\mathrm{d}f}{\mathrm{d}y}$ is a single symbol representing the derived function, it is *not* the ratio of two quantities $\mathrm{d}f$ and $\mathrm{d}y$.

Although there are systematic ways of finding derived functions from first principles, you will not be required to use them in this course. Indeed, physicists are rarely required to do this because tables of derivatives already exist for all the well-known functions, and derivatives of more complicated functions can usually be expressed as combinations of those basic derivatives. Table 1.6 lists a few of the basic derivatives along with the simplest of the rules for combining them — it also gives some explicit examples of functions and their derivatives. Computer packages are now available that implement the rules of differentiation, these are often used to determine the derivatives of more complicated functions (Figure 1.28).

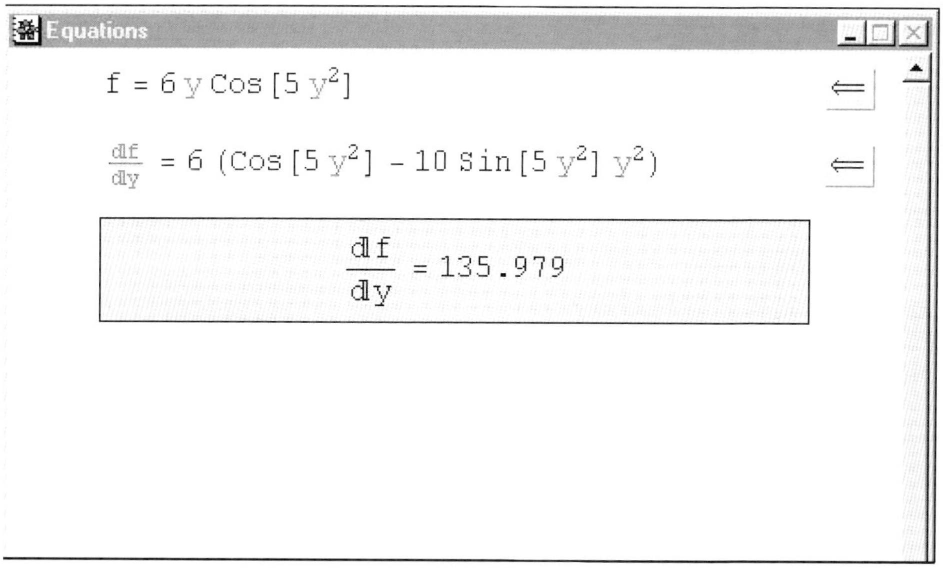

Figure 1.28 A function and its derivative, as displayed by an algebraic computing package.

Table 1.6 Some simple derivatives. The functions f, g and h depend on the variable y. The quantities A and n are constants, which may be positive, negative or zero. Note that n is not necessarily an integer.

Function $f(y)$	Derivative $\frac{df}{dy}$	Example
$f(y) = A$ (a constant)	$\frac{df}{dy} = 0$	$f(y) = 6$ $\frac{df}{dy} = 0$
$f(y) = y^n$ (a power of y)	$\frac{df}{dy} = ny^{n-1}$	$f(y) = y^3$ $\frac{df}{dy} = 3y^2$
$f(y) = Ay^n$ (a constant × a power of y)	$\frac{df}{dy} = nAy^{n-1}$	$f(y) = 2y^3$ $\frac{df}{dy} = 6y^2$
$f(y) = g(y) + h(y)$ (a sum of functions)	$\frac{df}{dy} = \frac{dg}{dy} + \frac{dh}{dy}$ (a sum of derivatives)	$f(y) = 2y^3 + 4y$ $\frac{df}{dy} = 6y^2 + 4$
$f(y) = Ag(y)$ (a constant × a function)	$\frac{df}{dy} = A\frac{dg}{dy}$ (a constant × a derivative)	$f(y) = 0.5 \times (2y^3 + 4y)$ $\frac{df}{dy} = 0.5 \times (6y^2 + 4)$

The idea of a derivative may be new to you and, if so, may seem rather strange. However, if you know the explicit form of a function, then there are several crucial advantages in using derivatives to determine gradients, rather than making measurements on a graph. In particular, looking up the derivative of a function in a table should be completely accurate, whereas measuring the gradient of the tangent to a graph is always approximate. For example, if $f(y) = y^2$ then the derivative of $f(y)$ is $\mathrm{d}f/\mathrm{d}y = 2y$ and evaluating the derivative at $y = 3$ to find the gradient at that particular value of y gives 6. This is an *exact* result that could not have been obtained with such precision from measurements on a graph. Moreover, if we want to know the gradient at many different values of y, all we need to do is to substitute each of those values into the general expression for the derivative, $\mathrm{d}f/\mathrm{d}y = 2y$. This is much simpler than drawing many different tangents and measuring their individual gradients.

4.4 Velocity and acceleration as derivatives

Recalling that the instantaneous velocity of a particle at time t is given by the gradient of its position–time graph at that time, we can now use the terminology of functions and derivatives to say that the velocity of the particle is given by the derivative of its position function. In terms of symbols:

$$v_x(t) = \frac{\mathrm{d}x(t)}{\mathrm{d}t}. \tag{1.14}$$

Similarly, we can say that the instantaneous acceleration of a particle is given by the derivative of the velocity function:

$$a_x(t) = \frac{\mathrm{d}v_x(t)}{\mathrm{d}t} \tag{1.15}$$

What's more we can use derivatives to simplify problems, as Example 1.1 shows.

Example 1.1

The position x of a particle at time t is given by the function $x(t) = kt^2$ where $k = -5\,\mathrm{m\,s^{-2}}$. Find (a) the velocity as a function of time; (b) the velocity at time $t = 3\,\mathrm{s}$.

Solution

(a) The velocity v_x is the derivative with respect to time of the position function $x(t)$ which is of the form At^n, with $A = k$ and $n = 2$. It therefore follows from the third of the standard results in Table 1.6 that

$$v_x(t) = \frac{\mathrm{d}x(t)}{\mathrm{d}t} = 2kt. \tag{1.16}$$

This is the required answer. No measuring of the gradients of tangents to curves is involved!

(b) Remembering that k is given as $-5\,\mathrm{m\,s^{-2}}$, the velocity at time $t = 3\,\mathrm{s}$ is now easily obtained from Equation 1.16, as follows

$$v_x(3\,\mathrm{s}) = 2kt = 2 \times (-5\,\mathrm{m\,s^{-2}}) \times (3\,\mathrm{s}) = -30\,\mathrm{m\,s^{-1}}.$$

Note that we have multiplied $\mathrm{m\,s^{-2}}$ by s to give $\mathrm{m\,s^{-1}}$, i.e. the units have been treated just like algebraic quantities.

Question 1.15 If the velocity $v_x(t)$ of a particle is given by $v_x(t) = kt^2$, where $k = 4\,\mathrm{m\,s^{-3}}$, find a general expression for the acceleration $a_x(t)$. What is the value of $a_x(3\,\mathrm{s})$? ■

Although it would be quite wrong to think of $\mathrm{d}x/\mathrm{d}t$ as a ratio of the quantities $\mathrm{d}x$ and $\mathrm{d}t$, it is useful to regard $\mathrm{d}x/\mathrm{d}t$ as consisting of an entity $\mathrm{d}/\mathrm{d}t$ that acts on the function $x(t)$. The entity $\mathrm{d}/\mathrm{d}t$ is a mathematical instruction to differentiate the function that follows, $x(t)$ in this case. It is an example of what a mathematician would call an **operator**. Adopting this view, we can say that if, for example, $x(t) = kt^2 + ct$, then

$$\frac{\mathrm{d}x}{\mathrm{d}t} = \frac{\mathrm{d}}{\mathrm{d}t}(kt^2 + ct).$$

Using the rule for differentiating the sum of two functions, from Table 1.6, we may write this as

$$\frac{\mathrm{d}x}{\mathrm{d}t} = \frac{\mathrm{d}}{\mathrm{d}t}(kt^2) + \frac{\mathrm{d}}{\mathrm{d}t}(ct).$$

Using the third result in Table 1.6 we may work out both the derivatives on the right-hand side to obtain

$$\frac{\mathrm{d}x}{\mathrm{d}t} = 2kt + c.$$

Regarding $\mathrm{d}/\mathrm{d}t$ as an operator also suggests another way of writing the acceleration. We already know that

$$a_x(t) = \frac{\mathrm{d}v_x}{\mathrm{d}t} \quad \text{and} \quad v_x(t) = \frac{\mathrm{d}x}{\mathrm{d}t}$$

so it seems sensible to write

$$a_x(t) = \frac{\mathrm{d}}{\mathrm{d}t}(v_x) = \frac{\mathrm{d}}{\mathrm{d}t}\left(\frac{\mathrm{d}x}{\mathrm{d}t}\right). \tag{1.17}$$

This emphasizes that the acceleration of a particle is the rate of change of the rate of change of the position, or if you prefer, the derivative of the derivative of the position function. Either of these formulations is a bit of a mouthful, so it is more conventional to refer to the acceleration as the **second derivative** of $x(t)$ and to represent it symbolically by

$$a_x(t) = \frac{\mathrm{d}^2x(t)}{\mathrm{d}t^2}. \tag{1.18}$$

Once again, it would be quite wrong to think of this as some kind of ratio of d^2x and $\mathrm{d}t^2$; it simply indicates that a particle's position function, $x(t)$, must be differentiated twice in order to find the particle's acceleration.

Note: If you are not already familiar with derivatives you should pay particular attention to the positioning of the superscripts in the second derivative symbol. Newcomers to differentiation often make the mistake of writing $\mathrm{d}x^2/\mathrm{d}t^2$ when they mean $\mathrm{d}^2x/\mathrm{d}t^2$. Remembering that it is the operator $\mathrm{d}/\mathrm{d}t$ that is to be squared rather than the function $x(t)$ may help you to avoid this error.

Example 1.2

Suppose, as in Example 1.1, the position of a particle is given by $x(t) = kt^2$ where k is a constant. Find $a_x(t)$.

Solution

We use Equation 1.14 and the third result in Table 1.6 with $A = k$, $y = t$ and $n = 2$, to obtain

$$v_x(t) = \frac{\mathrm{d}x(t)}{\mathrm{d}t} = 2kt. \qquad \text{(Eqn 1.16)}$$

It then follows from Equation 1.18, that

$$a_x(t) = \frac{\mathrm{d}^2x(t)}{\mathrm{d}t^2} = \frac{\mathrm{d}v_x(t)}{\mathrm{d}t} = 2k$$

where we have again used the third result in Table 1.6 but with $A = 2k$ and $n = 1$.

Question 1.16 Suppose that the vertical position x of a test vehicle falling down a drop-shaft is given by the quadratic function

$$x(t) = k_0 + k_1t + k_2t^2$$

where k_0, k_1 and k_2 are constants. Work out an expression for the vehicle's acceleration $a_x(t)$ in terms of k_0, k_1 and k_2, taking care to indicate each step in your working. ■

4.5 The signed area under a general velocity–time graph

We have already seen (in Section 3.6) that in the context of uniform motion, the signed area under a particle's velocity–time graph, between two given times, represents the change in the particle's position during that time interval, with a positive area corresponding to displacement in the positive direction. In the case of uniform motion, the velocity–time graph was a horizontal line and the area under the graph was rectangular.

Now, in the context of non-uniform motion, it seems natural to ask if the same interpretation can be given to the area under a general velocity–time graph, such as that between t_1 and t_2 in Figure 1.29. The answer to this is a definite yes, though a rigorous proof is beyond the scope of this book, so what follows is simply a plausibility argument.

Figure 1.29 The area under the velocity–time graph between t_1 and t_2 for an accelerating particle.

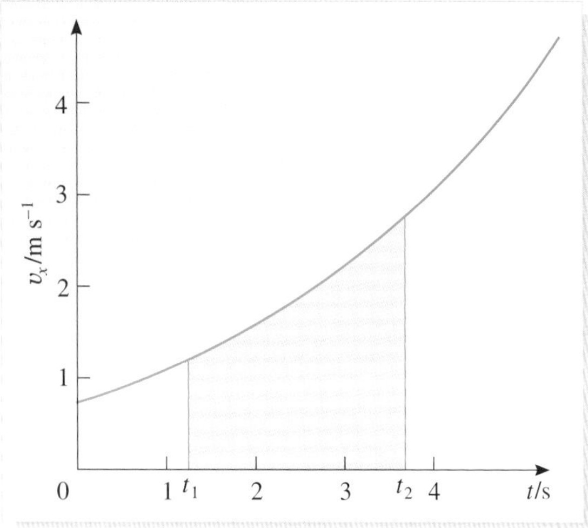

In Figure 1.29, the colour-shaded area under the graph does not take the shape of a rectangle. However, it may be approximately represented by a sum of rectangular areas, as indicated in Figure 1.30. To produce Figure 1.30 we have broken the time between t_1 and t_2 into a number of small intervals, each of identical duration Δt and within each small interval, the velocity has been approximated by a constant, which can be taken to be the average velocity during that interval. As a result, the area of each rectangular strip in Figure 1.30 represents the approximate change of position over a short time Δt and the sum of those areas represents the approximate change of position over the interval t_1 to t_2. Now, if we were to repeat this process while using a smaller value for Δt, as in Figure 1.31, then we would have more strips between t_1 and t_2; their total area, representing the approximate change in position between t_1 and t_2, would be an

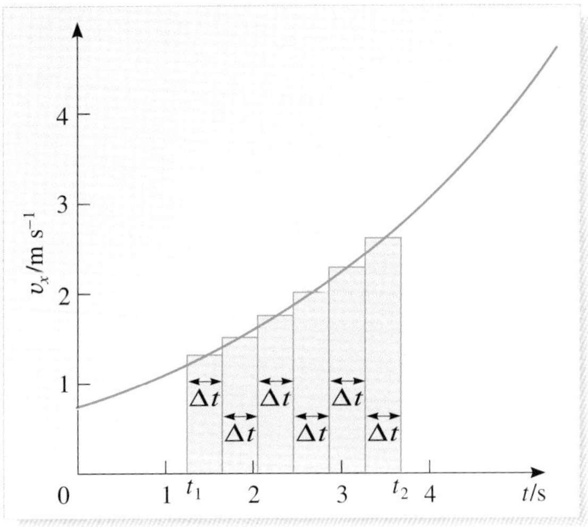

Figure 1.30 The area under the velocity–time graph of Figure 1.29, broken up into thin rectangular strips.

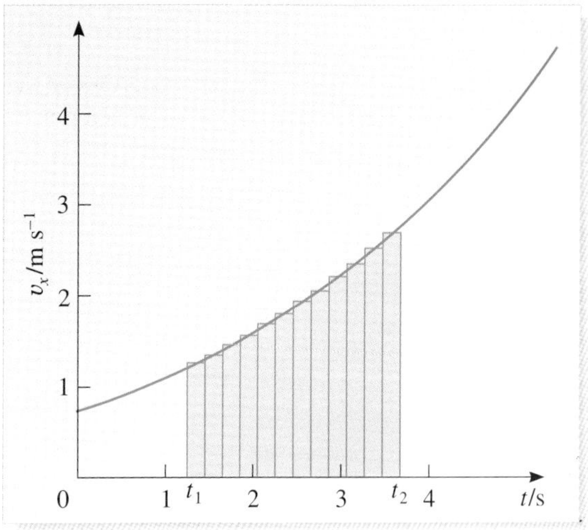

Figure 1.31 The area under the velocity–time graph of Figure 1.29, broken up into even more rectangular strips by reducing the value of Δt.

even closer approximation to the true area shown in Figure 1.29. Given these results, it seems reasonable to suppose that if we allowed Δt to become smaller and smaller, while the number of rectangular strips between t_1 and t_2 became correspondingly larger and larger, then we would eventually find that the area under the graph in Figure 1.29 was *exactly* equal to the change in position between t_1 and t_2.

This conclusion is in fact correct and can be proved in a rigorous way by considering what mathematicians call a *limit*, in this case 'the limit as Δt tends to zero'. We shall not pursue that here, but we should note that it wasn't until the early nineteenth century that the mathematics of limits was properly formulated, although it was in use long before then. It is also worth pointing out that it was the development of the idea of a limit that finally laid Zeno's paradox to rest. Just as the increasing number of diminishing strips can have a finite total area, so the increasing number of smaller steps that Achilles must take to reach the tortoise can have a finite sum and be completed in a finite time. Rigorous mathematical reasoning agrees with our everyday experience in telling us that motion can exist and that athletes outrun tortoises!

Question 1.17 Figure 1.32 shows the velocity–time graph for a particle with constant acceleration. (a) What is the displacement of the particle from its initial position after 6 s? (b) What is the distance travelled by the particle between $t = 2$ s and $t = 6$ s? (You may find it useful to know that the formula for the area of a triangle is: area = half the base × height, and that for the area of a trapezium is: area = base × (half the sum of the lengths of the parallel sides).) ■

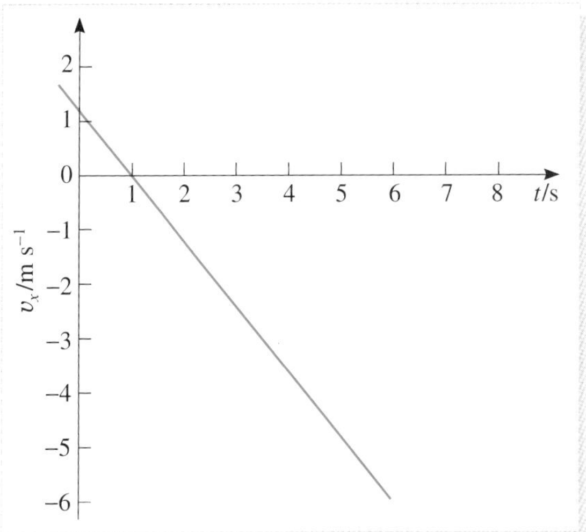

Figure 1.32 The velocity–time graph for Question 1.17.

5 Uniformly accelerated motion along a line

5.1 Describing uniformly accelerated motion

An important special case of non-uniform motion along a line is that which arises when an object is subjected to constant acceleration. This kind of motion is called **uniformly accelerated motion**. An object falling under gravity near to the surface of the Earth, such as the apple of Figure 1.20, provides an approximate realization of such motion. (Air resistance, which increases with speed, prevents the acceleration from being truly constant in such cases.) A more precise realization of uniformly accelerated motion is provided by an object falling under gravity close to the surface of an airless body such as the Moon (see Figure 1.33) or by a falling object in an evacuated (i.e. airless) drop-tower or drop-shaft of the kind discussed in Section 1 of this chapter.

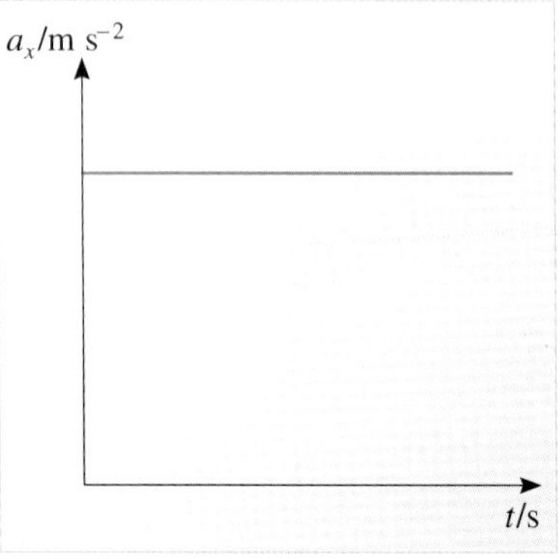

Figure 1.33 The acceleration–time graph for an object with constant (positive) acceleration.

The acceleration–time graph for a uniformly accelerated body is simple; it's just a horizontal line of the kind shown in Figure 1.33. The value of this constant acceleration (which may be positive or negative) represents the *gradient* of the velocity–time graph at any moment. It follows that the velocity–time graph must have the same gradient at all points and must therefore be a straight line of the kind shown in Figure 1.34. Note that the sign of the acceleration determines whether the velocity–time graph slopes up or down, and the value of the acceleration determines the precise value of the gradient. However, the acceleration does not determine the initial value of v_x so we have arbitrarily chosen a point on the v_x-axis to represent this value and labelled it u_x. From Figure 1.34, we deduce that if

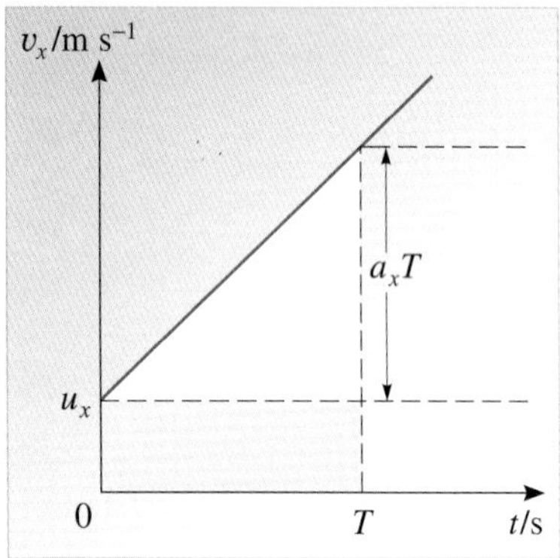

Figure 1.34 A velocity–time graph that is consistent with the acceleration–time graph of Figure 1.33. Note that the intercept has been chosen arbitrarily; only the gradient is determined by the acceleration.

$$a_x(t) = \text{constant} \tag{1.19}$$

then

$$v_x(t) = u_x + a_xt, \tag{1.20}$$

where a_x represents the *constant* value of $a_x(t)$.

It's rather more difficult to deduce the position–time graph that corresponds to uniformly accelerated motion, but the steady change of velocity with time certainly implies that the *gradient* of the position–time graph must also change steadily with time. In fact, given a velocity–time graph like that in Figure 1.34, the corresponding position–time graph will be of the general form shown in Figure 1.35.

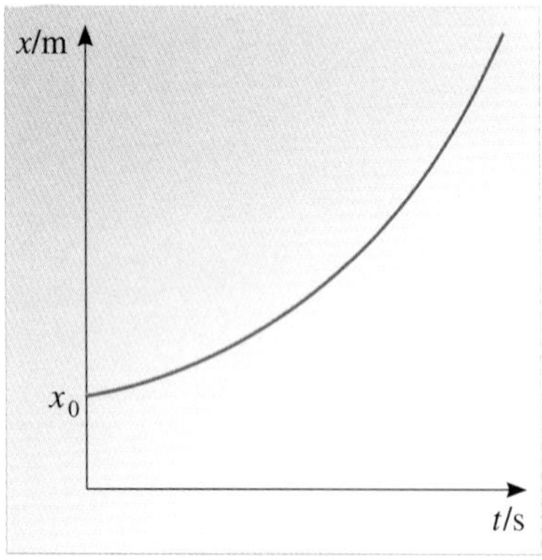

Figure 1.35 A position–time graph that is consistent with the acceleration–time graph of Figure 1.33 and the velocity–time graph of Figure 1.34. Note that the intercept has been chosen arbitrarily; only the gradient is determined by velocity.

Once again, the intercept with the vertical axis (representing the initial position at $t = 0$) is not determined by anything we have said so far; it has therefore been chosen arbitrarily and labelled x_0. The curve however is not arbitrary since the displacement from x_0 at any particular time $t = T$ is determined by the area under the velocity–time graph between $t = 0$ and $t = T$. That area will be the sum of two parts, a rectangle of height u_x and base length T, and a triangle of height a_xT and base length T. Since the area of a triangle is half the product of its height and its base length, it follows that the displacement from x_0 at time T will be

$$s_x(T) = u_xT + \tfrac{1}{2}a_xT^2$$

and the position of the uniformly accelerating particle, at any time t, will be $x(t) = x_0 + s_x(t)$, that is

$$x(t) = x_0 + u_xt + \tfrac{1}{2}a_xt^2. \tag{1.21}$$

Equations 1.19, 1.20 and 1.21 provide an essentially complete description of uniformly accelerated motion in one dimension and have many applications. They are not the most common form of the equations of uniform acceleration. We will discuss those in the next subsection, but before doing so let's use the method of differentiation to confirm the consistency of the equations we have deduced.

Starting from Equation 1.21, the right-hand side of which is a quadratic function of t, we expect to find that

$$v_x(t) = \frac{\mathrm{d}x(t)}{\mathrm{d}t} = \frac{\mathrm{d}}{\mathrm{d}t}\left(x_0 + u_x t + \tfrac{1}{2}a_x t^2\right).$$

Using the rules and results of Table 1.6 to carry out the differentiation (which is just like that in Question 1.16) we find that

$$v_x(t) = \frac{\mathrm{d}}{\mathrm{d}t}\left(x_0 + u_x t + \tfrac{1}{2}a_x t^2\right) = u_x + a_x t$$

in complete agreement with Equation 1.20. Similarly, differentiating the linear function that appears on the right-hand side of Equation 1.20, we expect to find

$$a_x(t) = \frac{\mathrm{d}v_x(t)}{\mathrm{d}t} = \frac{\mathrm{d}}{\mathrm{d}t}(u_x + a_x t).$$

Again, performing the differentiation, using Table 1.6, confirms our expectations:

$$a_x(t) = \frac{\mathrm{d}}{\mathrm{d}t}(u_x + a_x t) = a_x.$$

We see that in this case, $a_x(t)$ is just the constant acceleration a_x from which we started.

This short exercise in checking consistency gives just a hint of the immense power of differentiation to simplify a wide range of tasks and investigations.

Returning now to the description of uniformly accelerated motion, let's gather together our results so far:

$$x = x_0 + u_x t + \tfrac{1}{2}a_x t^2 \qquad (1.22)$$

$$v_x = u_x + a_x t \qquad (1.23)$$

$$a_x = \text{constant}. \qquad (1.24)$$

These are the equations we shall use, rearrange and extend in the next subsection.

5.2 The equations of uniformly accelerated motion

Equations 1.22, 1.23 and 1.24 provide a complete description of uniformly accelerated motion. By combining them appropriately, it is possible to solve a wide class of problems concerning the kinematics of uniformly accelerated motion. Nonetheless, those particular equations are not always the best starting point for the most common problems. For example, it is often the case that we want to know the displacement from the initial position after some specified period of constant acceleration, rather than the final position x. In such circumstances it is useful to subtract x_0 from both sides of Equation 1.22 and use the definition $s_x = x - x_0$ to write the resulting equation as

$$s_x = u_x t + \tfrac{1}{2}a_x t^2. \qquad (1.25)$$

More significantly, it is often the case that we need to find the final velocity v_x when all we are given is the (constant) acceleration a_x, the initial velocity u_x and the displacement s_x. The problem can be solved using Equations 1.23 and 1.25, but doing so involves finding the duration of the motion t, which is not required as part of the answer. It would be more convenient to use an equation that related v_x to a_x, u_x and s_x directly, thus avoiding the need to work out t altogether. Fortunately, it is possible to find such an equation by using a standard mathematical procedure called **elimination**.

The first step in the process is to identify a set of equations that contain the variables we want to relate, along with at least one variable we can eliminate. In this case we want to eliminate t from Equations 1.23 and 1.25. The second step usually involves rearranging one of the equations so that the unwanted variable is isolated on the left-hand side, thus becoming the **subject** of that equation. In this case, we can subtract u_x from both sides of Equation 1.23, divide both sides by a_x and then interchange the two sides to give

$$t = \frac{(v_x - u_x)}{a_x}. \tag{1.26}$$

Having obtained this relation from one of the equations, the third step is to use it to eliminate the unwanted variable from all the other equations. In our case this means replacing t by $(v_x - u_x)/a_x$ throughout Equation 1.25, so we obtain

$$\begin{aligned} s_x &= u_x \frac{(v_x - u_x)}{a_x} + \frac{a_x}{2} \frac{(v_x - u_x)^2}{a_x^2} \\ &= u_x \frac{(v_x - u_x)}{a_x} + \frac{a_x}{2} \frac{(v_x^2 - 2v_x u_x + u_x^2)}{a_x^2} \\ &= u_x \frac{(v_x - u_x)}{a_x} + \frac{1}{2} \frac{(v_x^2 - 2v_x u_x + u_x^2)}{a_x} \end{aligned}$$

so after a little algebra, we obtain

$$s_x = \frac{v_x^2 - u_x^2}{2a_x}$$

which can be rearranged to give the result

$$v_x^2 = u_x^2 + 2a_x s_x. \tag{1.27}$$

Equations 1.23, 1.25 and 1.27 are the most frequently used equations of uniformly accelerated motion and are usually referred to collectively as the constant acceleration equations (or the uniform acceleration equations).

Constant (or uniform) acceleration equations

$$s_x = u_x t + \tfrac{1}{2} a_x t^2 \tag{1.28a}$$

$$v_x = u_x + a_x t \tag{1.28b}$$

$$v_x^2 = u_x^2 + 2a_x s_x \tag{1.28c}$$

It follows from these equations that

$$s_x = \tfrac{1}{2}(v_x + u_x)t. \qquad (1.28d)$$

Remember, these are not universal equations that describe every form of motion. They apply *only* to situations in which the acceleration a_x is *constant*.

Question 1.18 Starting from the constant acceleration equations, use the elimination procedure to derive Equation 1.28d.

Question 1.19 Give a graphical interpretation of Equation 1.28d in terms of the area under a velocity–time graph for the case of uniformly accelerated motion.

Question 1.20 An object has a final velocity of $30.0\,\mathrm{m\,s^{-1}}$, after accelerating uniformly at $2.00\,\mathrm{m\,s^{-2}}$ over a displacement of 20.0 m. (a) What was the initial speed of the object? (b) For how long was the object accelerated?

Question 1.21 A train accelerates uniformly along a straight track at $2.00\,\mathrm{m\,s^{-2}}$ from an initial velocity of $4.00\,\mathrm{m\,s^{-1}}$ to a final velocity of $16.00\,\mathrm{m\,s^{-1}}$. What is the train's displacement from its initial position at the end of this interval? ■

5.3 The acceleration due to gravity

In the absence of air resistance, an object falling freely under the influence of the Earth's gravity, close to the surface of the Earth, experiences an acceleration of about $9.81\,\mathrm{m\,s^{-2}}$ in the downward direction. The precise value of the magnitude is indicated by the symbol g and varies slightly from place to place due to variations in surface altitude, the effect of the Earth's rotation and variations in the internal composition of the Earth. Some typical values for the **magnitude of the acceleration due to gravity**, g, at various points on the Earth's surface, are given in Table 1.7.

Table 1.7 The magnitude of the acceleration due to gravity at various points on the Earth's surface.

Location	$g/\mathrm{m\,s^{-2}}$
North Pole	9.83
London	9.81
New York	9.80
Equator	9.78
Sydney	9.80

Incidentally, one of the reasons that the value of g is well known across much of the Earth's surface is that extensive surveys have been carried out in which g has been accurately measured by timing the swing of very carefully constructed pendulums. The way in which g influences the oscillations of a simple pendulum is discussed in Chapter 3.

5.4 Drop-towers revisited

In Section 1 we described how research into near weightless conditions can be carried out on Earth by using a drop-tower or a drop-shaft (Figure 1.36). We are now in a position to examine drop-shafts in more detail (Example 1.3).

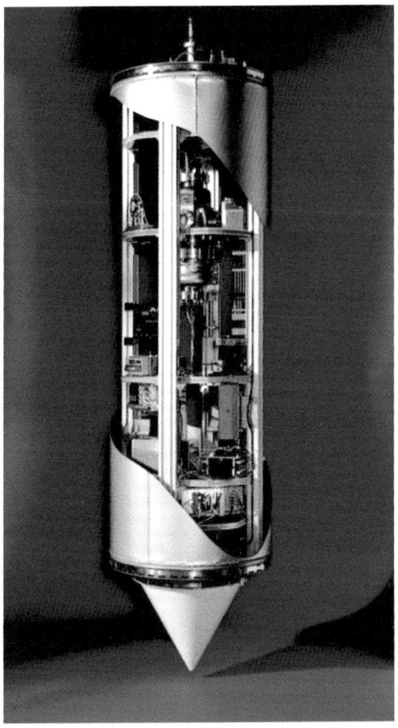

Figure 1.36 The test vehicle at the Bremen drop-tower.

Example 1.3

Consider a vertical shaft of total length 700 m, with free fall taking place for the first 500 m and constant deceleration for the final 200 m. Work out:

1 The time to fall the first 500 m. (This would be the time during which near weightless experiments could be done.)

2 The velocity at the end of the first 500 m.

3 The acceleration needed to reduce the velocity to zero at the bottom of the shaft.

Solution

It is convenient to split the calculations into three corresponding parts.

Part 1: Finding the time to fall the first 500 m When carrying out calculations it is always wise to choose a coordinate system that fits the problem. In this case it seems natural to choose an x-axis that points down the shaft, so that downward displacements and velocities will be positive and the acceleration due to gravity will be $a_x = +g$. It is also sensible to use the equations that lead to the least amount of work. In this case, we know only the initial velocity (assuming the object is initially at rest), the acceleration (g) and the final displacement (500 m). So, in order to work out the time to fall, the best approach is to use Equation 1.28a

$$s_x = u_x t + \tfrac{1}{2} a_x t^2 \qquad \text{(Eqn 1.28a)}$$

with $s_x = 500\,\text{m}$, $u_x = 0\,\text{m}\,\text{s}^{-1}$ and $a_x = g = 9.81\,\text{m}\,\text{s}^{-2}$. Setting $u_x = 0\,\text{m}\,\text{s}^{-1}$ and rearranging Equation 1.28a we obtain

$$t = \sqrt{\frac{2s_x}{a_x}} = \sqrt{\frac{2 \times 500\,\text{m}}{9.81\,\text{m}\,\text{s}^{-2}}}$$
$$= 10.1\,\text{s}.$$

Part 2: Finding the velocity at the end of the first 500 m To calculate the velocity after a free fall of 500 m, there are two convenient methods. We could either

(a) use $\quad v_x = u_x + a_x t \qquad$ (Eqn 1.28b)

with $u_x = 0\,\text{m}\,\text{s}^{-1}$, $a_x = g = 9.81\,\text{m}\,\text{s}^{-2}$ and $t = 10.1\,\text{s}$, or

(b) use $\quad v_x^2 = u_x^2 + 2a_x s_x \qquad$ (Eqn 1.28c)

with $u_x = 0\,\text{m}\,\text{s}^{-1}$, $a_x = g = 9.81\,\text{m}\,\text{s}^{-2}$ and $s_x = 500\,\text{m}$.

We choose the second method since it uses only values that we have been given rather than ones we have calculated. (It is always possible that we made some slip in our calculation.) So using Equation 1.28c with $u_x = 0\,\text{m}\,\text{s}^{-1}$, we obtain

$$v_x = \sqrt{2a_x s_x} = \sqrt{2 \times 9.81 \times 500}\,\text{m}\,\text{s}^{-1} = 99.0\,\text{m}\,\text{s}^{-1}.$$

Part 3: Finding the acceleration over the final 200 m To find the uniform acceleration necessary to bring the object to rest at the bottom of the shaft we can again make use of Equation 1.28c. This time, the quantity a_x is what we want to calculate and the known parameters are $v_x = 0\,\text{m}\,\text{s}^{-1}$, $u_x = 99.0\,\text{m}\,\text{s}^{-1}$ and $s_x = 200\,\text{m}$. Substituting $v_x = 0\,\text{m}\,\text{s}^{-1}$ and rearranging Equation 1.28c gives

$$a_x = -\frac{u_x^2}{2s_x} = -\frac{(99.0)^2}{2 \times 200}\,\text{m}\,\text{s}^{-2} = -24.5\,\text{m}\,\text{s}^{-2}.$$

Question 1.22 Consider the following proposal to roughly double the period of weightlessness in the 140 m Bremen drop-tower. The basic idea is that the drop-vehicle should be launched from the bottom of the tower with just the right velocity to enable it to reach the top of the tower with zero velocity. If the vehicle is uniformly accelerated over the first 10 m to give it the necessary upward velocity it can then rise freely for the remaining 130 m, pass through its highest point, and fall freely for another 130 m before being brought to rest over the final 10 m of its descent. Using a coordinate system in which the upward direction is positive, answer the following questions.

(a) What must be the launch velocity of the vehicle if it is to freely rise 130 m?

(b) What constant total acceleration must be applied over the first 10 m if the vehicle is to attain the required launch velocity? How long will this last?

(c) How long will the vehicle actually spend in free motion?

(d) What will be the acceleration of the vehicle when at its highest point?

(e) What will be the displacement of the vehicle from its starting point when it returns to that point at the end of the trip, and through what distance will it have travelled during the round trip?

(f) Sketch a rough acceleration–time graph for the entire motion, and briefly comment on the feasibility of the whole proposal. ■

6 Closing items

6.1 Chapter summary

1 A coordinate system provides a systematic means of specifying the position of a particle. A system in one dimension involves choosing an origin and a positive direction in which values of the position coordinate increase. Values of the position coordinate are positive or negative numbers multiplied by an appropriate unit of length, usually the SI unit of length, the metre (m).

2 The movement of a particle along a line can be described graphically by plotting values of the particle's position x, against the corresponding times t, to produce a position–time graph. Alternatively, by choosing an appropriate reference position x_{ref} and defining the displacement from that point by $s_x = x - x_{ref}$, the motion may be described by means of a displacement–time graph.

3 Uniform motion along a line is characterized by a straight-line position–time graph that may be described by the equation

$$x = v_x t + x_0 \qquad (1.6a)$$

where v_x and x_0 are constants. Physically, v_x represents the particle's velocity, the rate of change of its position with respect to time, and is determined by the gradient of the position–time graph

$$v_x = \frac{\Delta x}{\Delta t} = \frac{x_2 - x_1}{t_2 - t_1}. \qquad (1.5)$$

x_0 represents the particle's initial position, its position at $t = 0$, and is determined by the intercept of the position–time graph, the value of x at which the plotted line crosses the x-axis, provided that axis has been drawn through $t = 0$.

4 Non-uniform motion along a line is characterized by a position–time graph that is not a straight line. In such circumstances the rate of change of position with respect to time may vary from moment to moment and defines the instantaneous velocity. Its value at any particular time is determined by the gradient of the tangent to the position–time graph at that time.

5 More generally, if the position of a particle varies with time in the way described by a function $x(t)$, then the way in which the (instantaneous) velocity varies with time will be described by the associated derived function or derivative

$$v_x(t) = \frac{\mathrm{d}x(t)}{\mathrm{d}t}. \qquad (1.14)$$

6 The instantaneous acceleration is the rate of change of the instantaneous velocity with respect to time. Its value at any time is determined by the gradient of the tangent to the velocity–time graph at that time. More generally, the way in which the (instantaneous) acceleration varies with time will be described by the derivative of the function that describes the instantaneous velocity, or, equivalently, the second derivative of the function that describes the position:

$$a_x(t) = \frac{\mathrm{d}v_x(t)}{\mathrm{d}t} = \frac{\mathrm{d}^2x(t)}{\mathrm{d}t^2}. \qquad (1.18)$$

7 Results and rules relating to differentiation and the determination of derivatives are contained in Table 1.6. The derivative of a constant is zero, the derivative of $f(y) = Ay^n$ is $\mathrm{d}f/\mathrm{d}y = nAy^{n-1}$.

8 The signed area under a velocity–time graph, between specified values of time, represents the change in position of the particle during that interval.

9 Uniformly accelerated motion is a special case of non-uniform motion characterized by a constant value of the acceleration, a_x = constant. In such circumstances the velocity is a linear function of time ($v_x(t) = u_x + a_xt$), and the position is a quadratic function of time ($x(t) = x_0 + u_xt + \frac{1}{2}a_xt^2$).

10 The most widely used equations describing uniformly accelerated motion are

$$s_x = u_xt + \tfrac{1}{2}a_xt^2 \qquad (1.28a)$$

$$v_x = u_x + a_xt \qquad (1.28b)$$

$$v_x^2 = u_x^2 + 2a_xs_x \qquad (1.28c)$$

$$s_x = \tfrac{1}{2}(v_x + u_x)t. \qquad (1.28d)$$

11 Position x, displacement s_x, velocity v_x, and acceleration, a_x, are all signed quantities that may be positive or negative, depending on the associated direction. The magnitude of each of these quantities is a positive quantity that is devoid of directional information. The magnitude of the displacement of one point from another, $s = |s_x|$, represents the distance between those two points, while the magnitude of a particle's velocity, $v = |v_x|$, represents the speed of the particle. The magnitude of the acceleration due to gravity is represented by the symbol g, and has the approximate value 9.81 m s^{-2} across much of the Earth's surface.

6.2 Achievements

Now that you have completed this chapter, you should be able to:

A1 Explain the meaning of all the newly defined (emboldened) terms introduced in this chapter.

A2 Draw, analyse and interpret position–time, displacement–time, velocity–time and acceleration–time graphs. Where appropriate, you should also be able to relate those graphs one to another and to the functions or equations that describe them, particularly in the case of straight-line graphs.

A3 Find the derivatives of simple polynomial functions, express physical rates of change as derivatives, and relate derivatives to the gradients of appropriate graphs.

A4 Solve simple problems involving uniform motion and uniformly accelerated motion by using appropriate equations. You should also be able to rearrange simple equations, to change the subject of an equation, and to eliminate variables between sets of equations.

A5 Describe the nature and purpose of drop-towers and drop-shafts, with particular reference to their role in simulating the near weightless conditions of space.

6.3 End-of-chapter questions

Question 1.23 Table 1.8 shows the atmospheric pressure P in pascals (Pa) at various heights h above the Earth's surface. Plot a graph to give a visual representation of the data in the table. Be careful to label your axes correctly. Explain why you have chosen to plot particular variables on the horizontal and vertical axes. Use your graph to find the rate of change of atmospheric pressure with height at $h = 10$ km.

Table 1.8 Data for Question 1.23.

h/km	P/Pa
0	101 325
5	48 586
10	23 297
15	11 171
20	5357
25	2569
30	1232

Question 1.24 (a) Define the terms *position* and *displacement*, and carefully distinguish between them. (b) The position x of a particle at time t is given in Table 1.9. Plot a position–time graph for the data in Table 1.9. (c) Using your graph, measure the velocity of the particle at $t = 5$ s. (d) A second particle undergoes uniform motion and has the same position and velocity as the first particle at time $t = 5$ s. What is the displacement of the second particle from the first at time $t = 10$ s?

Table 1.9 Data for Question 1.24.

t/s	x/km
0	0
1	0.443
2	0.984
3	1.64
4	2.45
5	3.44
6	4.64
7	6.11
8	7.91
9	10.1
10	12.8

Question 1.25 (a) Define the terms *velocity* and *acceleration*.

(b) The velocity v_x of a particle moving along the x-axis at various times t is given in Table 1.10. (i) Assuming the particle has a constant acceleration between the given positions, draw a velocity–time graph representing the data in Table 1.10. (ii) Use your graph to calculate the total displacement of the particle over the time interval given in the table.

Table 1.10 Data for Question 1.25.

v_x/m s^{-1}	t/s
4	0
−4	10
0	15
2	20

Comment In Questions 1.26 to 1.30, you are not required to draw any graphs.

Question 1.26 The variable z is related to the variable y by the equation

$$z = 3 + 2y + y^3.$$

(a) Find the derivative dz/dy. (b) Evaluate dz/dy at $y = 2$. (c) What would be the gradient of a graph of z plotted against y for the value $y = 2$?

Question 1.27 A rocket travels vertically away from the surface of the Moon. It is still close to the Moon's surface when it jettisons an empty fuel tank. The fuel tank initially travels with the same velocity as the rocket, but falls back to the Moon, reaching the Moon's surface 50 s after being released. If the fuel tank hits the surface at a speed of 50 m s^{-1}, calculate the speed of the fuel tank when it was jettisoned. You may assume that the magnitude of the acceleration due to gravity near the Moon's surface is 1.6 m s^{-2}.

Question 1.28 For time t greater than or equal to zero, a particle's position as it travels along the x-axis is described by the function $x(t) = At^2$, where $A = 4.0\,\mathrm{m\,s^{-2}}$. Use differentiation to calculate how fast the particle is travelling at $t = 10\,\mathrm{s}$.

Question 1.29 A vase falls to the ground from a shelf at height 1.80 m above the floor. Neglecting air resistance, calculate the time taken for the vase to strike the floor.

Question 1.30 (This question is more difficult than its predecessors.) A rocket is initially at rest on the Earth's surface. At time $t = 0\,\mathrm{s}$ the rocket motor is fired and the rocket accelerates vertically upwards at a constant $2.00\,\mathrm{m\,s^{-2}}$. After a time interval of 20.0 s the motor fails completely and the rocket accelerates back to Earth with a downward acceleration of magnitude $9.81\,\mathrm{m\,s^{-2}}$.

(a) Calculate the height reached by the rocket at the instant the motor fails.

(b) Calculate the velocity of the rocket at the instant the motor fails.

(c) Find an equation that, for situations involving constant acceleration, gives the displacement in terms of the initial velocity, the final velocity, and the constant acceleration.

(d) Calculate the distance travelled by the rocket from the instant the motor fails until the rocket reaches its highest point. Hence, find the total height gained by the rocket. [*Hint*: The equation you obtained in part (c) may be useful here.]

(e) Find the total time taken for the rocket to fall back to Earth from its highest point. ■

Chapter 2 Motion in a plane and in space

1 Long jumping — an example of motion in a plane

Long jumping is an ancient sport; it was already part of the Olympic games more than two and a half thousand years ago. The challenge that faces a long jumper (Figure 2.1), whether a novice or an Olympian, is quite simple to state; at the end of a short horizontal approach run, jump upwards before reaching a specified mark and land as far as possible from that mark. There are, of course, important points of technique, such as not falling backwards upon landing, that the novice must learn, but the essential problem is for the athlete to launch his or her body into the air in such a way that it travels the maximum possible distance before returning to the ground. How is this to be achieved? Should the jumper put more effort into gaining horizontal speed during the run-up or into gaining height and hence time in the air from the upward jump? Is there an optimal angle of launch that the long jumper should try for, and if so what is it? What is the physical limit to the length that might be attained, and how close are current day long jumpers to reaching that limit? These are exactly the sort of questions that physics can answer, and in this chapter you will learn how to work out those answers.

Figure 2.1 A long jumper in action.

The scientific process of modelling long jumping — of turning it into a mathematical problem, solving that problem, and then interpreting the results in a way that might help a real long jumper — is typical of the kind of process that arises in many areas of physics. You have already met some of the principles of modelling in Chapter 1, particularly the need to simplify the system under study so that, initially at least, the problem to be solved is not too great. We shall make similar simplifications here, by initially ignoring arms and legs and treating the long jumper as a particle. However in this case we cannot treat the jumper as moving solely along a line in one dimension. Long jumping necessarily involves a combination of horizontal and vertical motion and hence two-dimensional motion — motion in a plane. As you will see in this chapter, motion in two or more dimensions — motion in a plane or in space — is conventionally described in terms of quantities known as *vectors*. Learning how to use vectors will be another important outcome of this chapter.

2 Position and displacement in a plane

2.1 From one dimension to two

In Chapter 1, you saw how a one-dimensional coordinate axis can be defined and used to describe the position of a particle moving along a straight line. However, most motions are not confined to a straight line. Figure 2.2 is a case in point. It is a **stroboscopic photograph** of a ball. The ball was thrown into the air at a steep angle to the horizontal and its subsequent motion was captured on a single photograph by illuminating the ball with a rapid sequence of short-duration flashes, each separated by an equal amount of time from its successor. The fact that the path traced out by the ball is a curve in the vertical plane shows that the motion is two-dimensional. The fact that the ball moves different distances between successive flashes shows that its speed is changing with time. You have already learnt how to describe straight-line motion in terms of the position, velocity and acceleration at each instant. One of the aims of this chapter is to show you how to describe motion in a plane in a similar way.

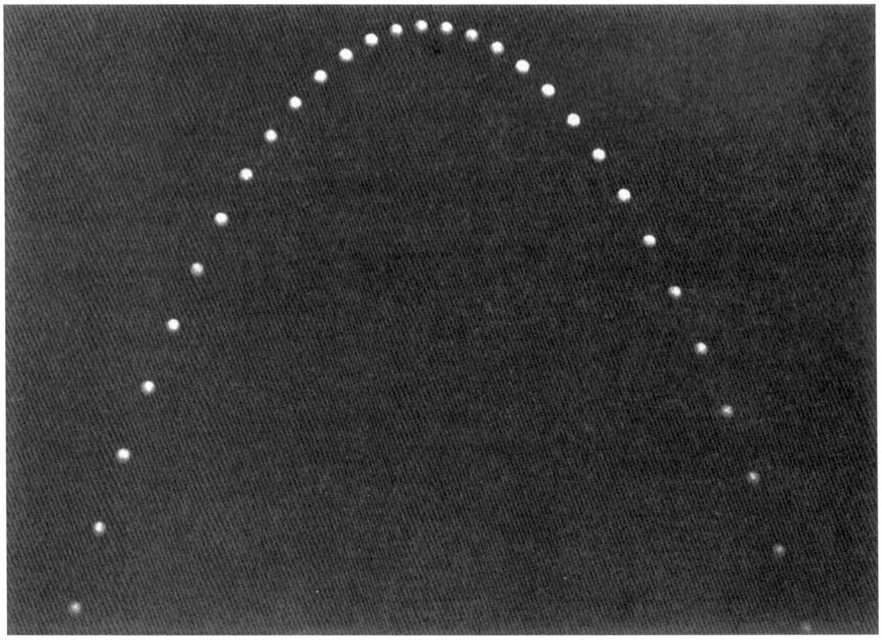

Figure 2.2 A stroboscopic photograph of a ball travelling in a vertical plane.

For the sake of simplicity we shall ignore the radius of the ball and the possibility that it might be rotating. We are therefore modelling the ball as a particle located at the ball's centre. This is a sufficiently detailed model for the purpose of learning how to describe motion in two dimensions.

2.2 Position and position vectors

Position coordinates in a plane

One way of specifying the position of a ball as it moves in two dimensions, is to use a pair of coordinate axes, located in the plane of the ball's motion. Figure 2.3 shows the simplest way of arranging such a pair of axes. In this case the axes meet at right angles (i.e. they are *orthogonal* or *mutually perpendicular*), and the single point at which they meet is taken to be the *origin* of coordinates measured along each of the axes. When using such an arrangement it is conventional to call the horizontal axis the x-axis and the vertical axis the y-axis. Both axes may be calibrated in metres, starting from zero at the origin and increasing in the direction of the arrows. The motion of the ball is then said to be confined to the xy-plane defined by these axes.

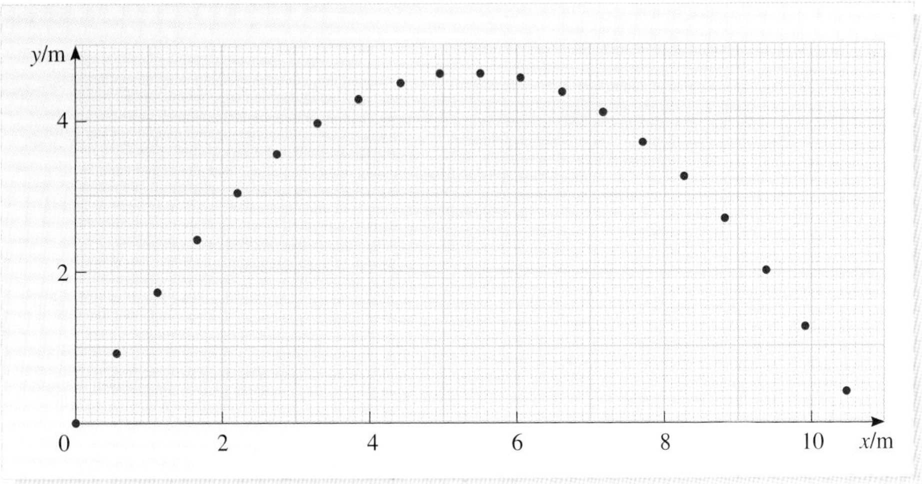

Figure 2.3 The position of a ball moving in a vertical plane can be specified by the x- and y-coordinates of its centre at any time. In this case the ball was launched at an angle of 60°, and the time interval between the successive positions shown was 0.1 s. Note that 1 cm on this diagram corresponds to 1 m in real space.

Using the two-dimensional coordinate system shown in Figure 2.3 the position of the ball at any time during its flight can be specified by quoting the values of its x- and y-coordinates at that time. The coordinate values may be determined by drawing perpendicular lines from the axes to the instantaneous position of the ball and noting the values at which those perpendicular lines meet the axes. Thus, as shown in Figure 2.4, the x- and y-coordinates of the point marked A are

$$x = 7.7\text{ m} \quad \text{and} \quad y = 3.7\text{ m}.$$

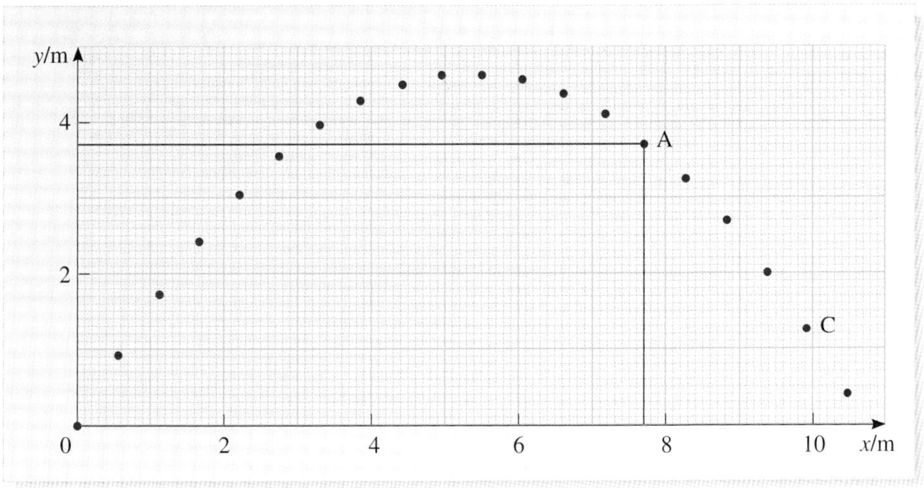

Figure 2.4 The coordinate values that determine the location of the ball at any time are found by drawing perpendicular lines from each axis to the position of the ball's centre.

It is conventional to indicate pairs of coordinate values such as $x = 7.7$ m and $y = 3.7$ m by writing them in parentheses, with a comma between the two values. Using this convention we can write the coordinates of point A as

$$(7.7 \text{ m}, 3.7 \text{ m}).$$

Note that this convention assumes that the coordinates will *always* be presented in the order x then y, otherwise it would be ambiguous. For this reason, pairs of coordinates presented in the form (x, y) are said to constitute an **ordered pair**.

Question 2.1 Specify the x- and y-coordinates of the ball when it has reached point C in Figure 2.4. Present your answer as an ordered pair. ■

All that has been said so far about the use of coordinates in Figures 2.3 and 2.4 should have been fairly familiar to you. It's really no different from the way in which coordinates were used in the last chapter to plot graphs. Even Figure 2.4's use of the axis label x/m, to show that the numbers along the horizontal axis represent values of x measured in units of metres, is a convention that you met earlier. However, there is another way of describing positions that may be less familiar but is even more useful in physics. This is the subject to which we now turn.

Position vectors in a plane

As Figure 2.5 shows, an ordered pair of coordinates such as (7.7 m, 3.7 m) determines a point A in a plane and hence an arrow from the origin to that point. Position-fixing arrows of this kind are very important in physics; they provide a pictorial representation of a quantity known as a **position vector** that can be used to describe the location of any point relative to the origin of a coordinate system.

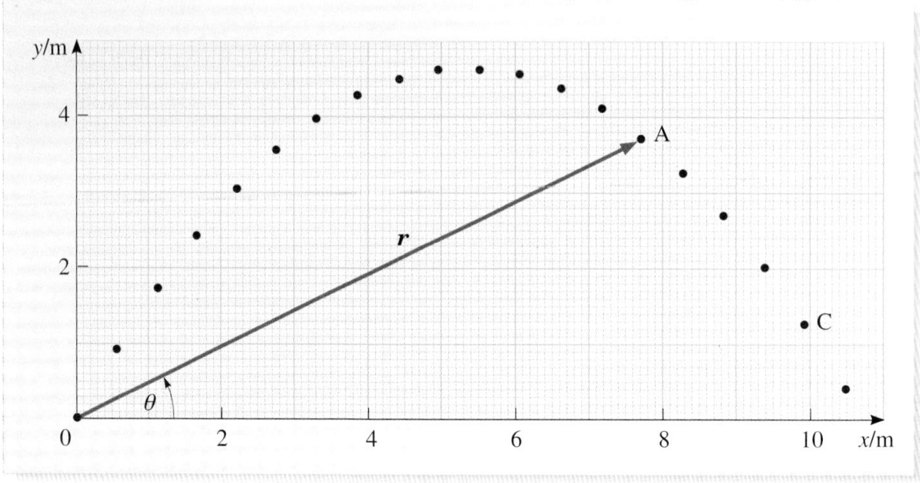

Figure 2.5 The position vector $\boldsymbol{r}$ of the point A is represented by the arrow from the origin to A.

Like the arrow used to represent it, the position vector of a point has two essential properties:

1 A length (usually referred to as the **magnitude** of the position vector) that records the distance from the origin to the point in question. This quantity must be measured in units of length and, since it represents a distance, can never be negative.

2 An orientation (usually referred to as the **direction** of the position vector) that records the angle between the position vector and some specified line, usually the x-axis.

By convention, any symbol representing a position vector in printed text (such as the $\boldsymbol{r}$ in Figure 2.5) is set in bold italic type. The magnitude of the vector can then be represented by the same letter, but without any emboldening (just a plain r would suffice for the case in Figure 2.5). There is no particular notational convention regarding the direction of a position vector, but it is often expressed in terms of the angle θ measured in the anticlockwise direction from the x-axis to the position vector (this too is indicated in Figure 2.5).

Question 2.2 Figure 2.5 is a scale drawing in which 1 cm represents 1 m. Estimate the magnitude and direction of the position vector $\boldsymbol{r}$ shown in the figure. Present your answer by giving (approximate) values for the quantities r and θ. ■

Using boldface type to represent printed position vectors is a common convention throughout physics. However, when using a pen or pencil, it is not easy to write bold symbols, so a different convention is needed in these circumstances. The most widely used convention is to write the position vector with a curly line underneath, as in $\underset{\sim}{\mathrm{r}}$. (This is consistent with the use of a wavy underline by editors and printers to indicate the instruction 'make bold'.) An alternative convention is to use straight underlining, or to draw a small arrow above the symbol, but we would encourage you to use the curly underline. The magnitude of a position vector is always written without any underlining, since it's just a simple distance.

In handwritten work a position vector may be written $\underset{\sim}{\mathrm{r}}$ and its magnitude is r or $|\underset{\sim}{\mathrm{r}}|$.

You *must* make sure that you underline all vectors in your written work; if you forget to do this in assignments and examinations you will almost certainly lose marks. You have been warned!

All this emphasis on whether a symbol should be bold or not may seem excessive, but it is important to realize that position vectors and their magnitudes are quite different mathematically, so they must be distinguished if errors are to be avoided in calculations. The magnitude of a position vector is simply a distance and as such may be written as a positive number times a suitable unit of measurement, such as a metre. The position vector itself, however, provides more information; it certainly has a magnitude, but it also has a direction. In fact, the position vector $\boldsymbol{r}$ of a point conveys the same information as the ordered pair of coordinates that also determines the location of the point. The two methods of describing position are equivalent. Thus, in the case of the position vector shown in Figure 2.5 we can write

$$\boldsymbol{r} = (7.7\,\mathrm{m}, 3.7\,\mathrm{m}).$$

More generally, the position vector $\boldsymbol{r}$ of any point with coordinates (x, y) may be specified by writing

$$\boldsymbol{r} = (x, y). \tag{2.1}$$

When a position vector is written in this way the coordinate values x and y are referred to as the **components** of the position vector. The position vector shown in Figure 2.5 has an x-component of 7.7 m and a y-component of 3.7 m. Note that once again we are dealing with an ordered pair and that care must be taken to preserve the ordering. The position vector (3.7 m, 7.7 m) is *not* the same as the position vector (7.7 m, 3.7 m)! Also, notice that each component consists of a number multiplied by a unit of length. It would be incorrect to state that $\boldsymbol{r} = (7.7, 3.7)$ with no mention of units, because each component is a physical length, characterized by both a number *and* a unit of measurement. However, the units can be placed outside the bracket if you prefer, as in $\boldsymbol{r} = (7.7, 3.7)\,\mathrm{m}$.

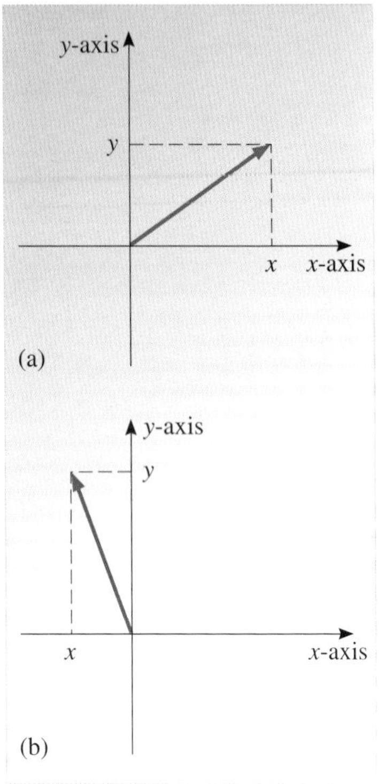

Figure 2.6 Any position vector in the xy-plane is completely specified by its x- and y-components. In case (a), the x- and y-components are both positive. In case (b), the x-component is negative and the y-component positive.

The components of a position vector may be positive or negative. Figure 2.6a shows a position vector that has a positive x-component and a positive y-component. In contrast, Figure 2.6b shows a position vector that has a negative x-component but a positive y-component. By choosing suitable combinations of positive and negative values for the components it is possible to write down the position vector of any point in the xy-plane. The origin itself is represented by the position vector $\boldsymbol{r} = (0, 0)$, which is known as the **zero vector** and is usually represented by the bold symbol **0** or $\underset{\sim}{0}$ in handwritten text.

Since a given position vector $\boldsymbol{r}$ may be expressed either in terms of its magnitude and direction (i.e. in terms of r and θ) or in terms of its components (i.e. in terms of x and y) it must be possible to relate r and θ to x and y. In order to do this you will need to be aware of some basic facts about the trigonometry of right-angled triangles. These essential facts are listed in Box 2.1.

Box 2.1 Triangles and trigonometry

Trigonometry is concerned with right-angled triangles such as that shown in Figure 2.7, and with the ratios of the lengths of the sides of such triangles. If the lengths of the two shorter sides opposite and adjacent to the angle θ are respectively denoted b and a, then the length c of the longest side, known as the hypotenuse, is given by **Pythagoras' theorem**, according to which

$$c^2 = a^2 + b^2 \quad \text{or} \quad c = \sqrt{a^2 + b^2}.$$

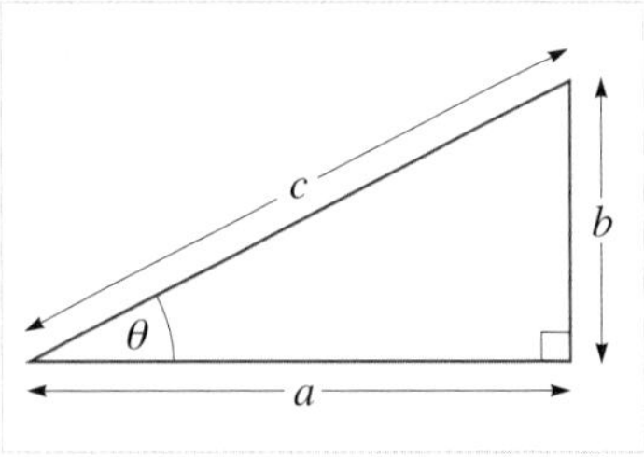

Figure 2.7 A right-angled (90°) triangle. The side opposite the right angle is called the hypotenuse and is of length c.

In addition, the ratios of the lengths a, b and c define the so called **trigonometric ratios**, provided that angle θ is less than 90°. The most important of the trigonometric ratios are called sine, cosine and tangent. They are usually abbreviated to sin, cos and tan and are defined as follows:

$$\sin\theta = b/c$$

$$\cos\theta = a/c$$

$$\tan\theta = b/a.$$

A very useful result that follows directly from Pythagoras' theorem is that

$$(\sin\theta)^2 + (\cos\theta)^2 = 1.$$

This is called an *identity* since it is true for all values of θ, and is usually written in the form

$$\sin^2\theta + \cos^2\theta = 1.$$

Note the positioning of the superscripts in this latter form. The square of $\sin\theta$ or $\cos\theta$ is often written in this way, but NEVER as $\sin\theta^2$ or $\cos\theta^2$ since this would be confused with the sine or cosine of θ^2.

Using your calculator you should be able to confirm that

$\sin 0° = 0 \quad \cos 0° = 1 \quad \tan 0° = 0$

$\sin 30° = \frac{1}{2} \quad \cos 30° = \frac{\sqrt{3}}{2} \quad \tan 30° = \frac{1}{\sqrt{3}}$

$\sin 45° = \frac{1}{\sqrt{2}} \quad \cos 45° = \frac{1}{\sqrt{2}} \quad \tan 45° = 1$

$\sin 60° = \frac{\sqrt{3}}{2} \quad \cos 60° = \frac{1}{2} \quad \tan 60° = \sqrt{3}$

$\sin 90° = 1 \quad \cos 90° = 0.$

These particular results are probably worth committing to memory, but if you do forget them you can easily recover them as decimals from your calculator, provided you set it to degree mode.

Note that tan 90° is not defined, since no triangle can have two right angles. If you try to find tan 90° you will get an error message.

In the case of the position vector $\boldsymbol{r}$ shown in Figure 2.8, it follows from Pythagoras' theorem that

$$r = \sqrt{x^2 + y^2}\,. \tag{2.2}$$

It also follows from the definition of the trigonometric ratios that

$$\cos\theta = \frac{x}{r} \tag{2.3a}$$

$$\sin\theta = \frac{y}{r}. \tag{2.3b}$$

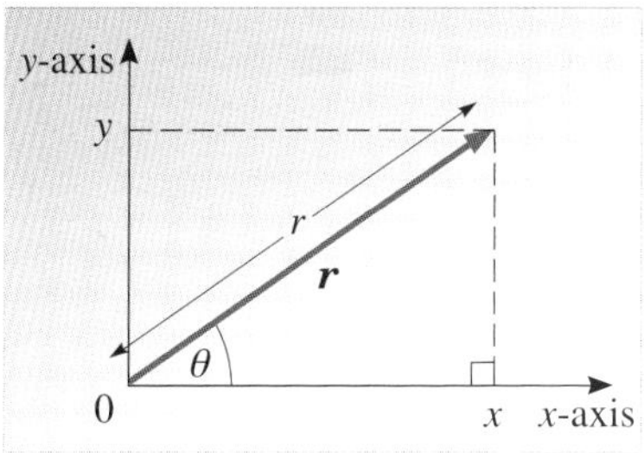

Figure 2.8 A position vector $\boldsymbol{r}$ of magnitude r and direction θ, with components x and y.

Given the value of sin θ (or cos θ) you can determine the value of θ by entering the given value into your calculator and the pressing the arcsin (or arccos) key. (On some calculators you have to press the key before entering the value.) Note that arcsin and arccos are sometimes referred to as '$\sin^{-1}$' and '$\cos^{-1}$', respectively.

Rearranging Equations 2.3a and 2.3b (by multiplying both sides of each equation by r) we find

$$x = r\cos\theta \tag{2.4a}$$

$$y = r\sin\theta. \tag{2.4b}$$

Using Equations 2.2 and 2.3, it is possible to work out the magnitude and direction of any position vector provided its components x and y are known. Using Equations 2.4a and 2.4b it is possible to work out the components if the magnitude and direction are known.

Question 2.3 Tackle Question 2.2 again by determining the x- and y-components from Figure 2.5 and then using Equations 2.2 and 2.3. ■

Figure 2.8 refers explicitly to the case where the x- and y-components are both positive and θ is in the range 0° to 90°. Nevertheless, Equations 2.2, 2.3 and 2.4 may be applied quite generally, whatever the signs of x and y, and irrespective of the value of θ. This is possible because the trigonometric ratios discussed above may be generalized to give a set of *trigonometric functions* that are defined for all values of θ, and it is these that are programmed into your calculator. The trigonometric

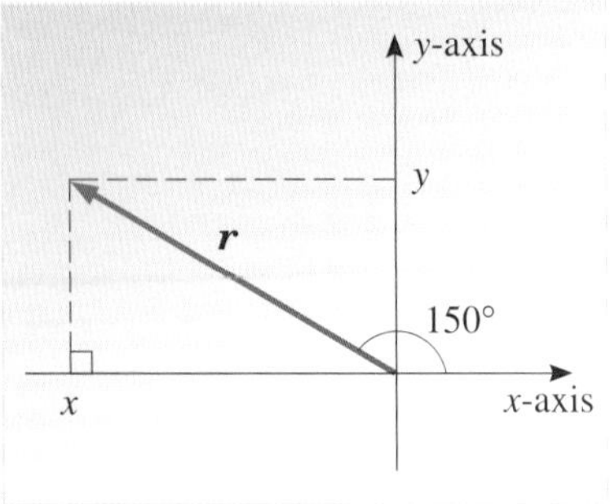

Figure 2.9 The x- and y-components of a position vector can be found using a calculator, even if θ is greater than 90°.

functions will be discussed in more detail in the next chapter. For the moment just accept the output of your calculator as far as sines and cosines are concerned (provided it's set to degree mode), but don't make similar assumptions about the arcsin (or $\sin^{-1}$) and arccos (or $\cos^{-1}$) keys.

Figure 2.9 shows a case where the x-coordinate is negative, and $\theta = 150°$. Even in this case Equations 2.3a and 2.3b still apply, giving

$$x = r\cos 150° \quad \text{and} \quad y = r\sin 150°.$$

The sine and cosine of 150° are easily found on a calculator, giving

$$x = -0.87r \quad \text{and} \quad y = 0.5r.$$

As you can see, the x-component is negative in this case as you would expect from Figure 2.9.

Question 2.4 Sketch very roughly (not on graph paper) an x-axis and a y-axis, meeting at a common origin, and show the following position vectors: $\boldsymbol{r}_A = (1\,\text{m}, 2\,\text{m})$, $\boldsymbol{r}_B = (1\,\text{m}, -2\,\text{m})$, $\boldsymbol{r}_C = (-1\,\text{m}, 2\,\text{m})$. What is the magnitude of each of these vectors, and what is its physical significance in each case? ■

Here is a summary of the main points to remember about position vectors.

The equation $\theta = \arctan(y/x)$ is only guaranteed to give the right value for θ when $0 \leq \theta \leq 90°$.

A position vector is a quantity with magnitude and direction that fixes the location of a point relative to an agreed origin. It may be specified in terms of its components along orthogonal axes which meet at the origin. The components are then equal to the coordinates of the specified point referred to those same orthogonal axes. If the position vector is printed as $\boldsymbol{r}$ then it may be handwritten as $\underset{\sim}{\text{r}}$. The magnitude of the position vector $\boldsymbol{r}$ is printed as r and handwritten as r; it represents a distance and can never be negative. The direction of the position vector may be expressed in terms of the angle θ measured anticlockwise from the (positive) x-axis to the position vector. In two dimensions, if $\boldsymbol{r} = (x, y)$ then $r = \sqrt{x^2 + y^2}$ and $\theta = \arctan(y/x)$.

2.3 Displacement and displacement vectors

You saw in Chapter 1 that a particle moving along the x-axis from position x_1 to position x_2 undergoes a displacement $s_x = x_2 - x_1$. We will now extend this idea to cover motion in two dimensions. Figure 2.10 represents the path of a particle, moving in the plane of the page, that passes through point 1, and then travels on through point 2. The **displacement vector** from point 1 to point 2 describes the distance and the direction from point 1 to point 2, and is represented in Figure 2.10 by the length and direction of the arrow whose tail is at point 1 and whose head is at point 2. The displacement vector is often given the bold symbol $\boldsymbol{s}$. (When writing the displacement by hand, use a curly underline, as in $\underset{\sim}{\text{s}}$.)

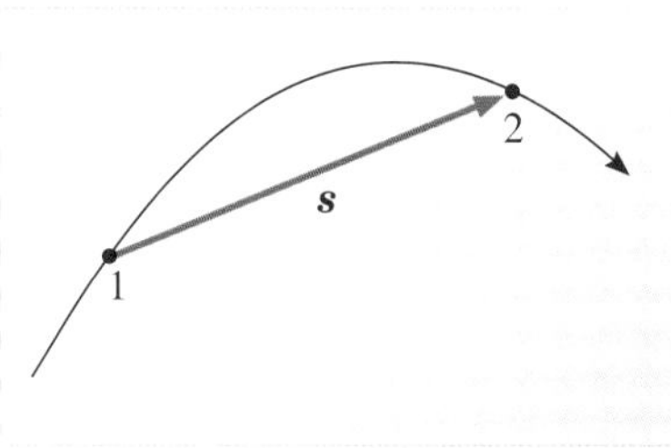

Figure 2.10 The displacement vector $\boldsymbol{s}$ from point 1 to point 2.

A displacement vector $\boldsymbol{s}$ describes a net change in position, so the symbol $\Delta\boldsymbol{r}$ is also used to represent it, where the Greek letter delta, Δ, indicates *change of*, as in Chapter 1. Since the relevant change of position in Figure 2.10 is $\Delta\boldsymbol{r} = \boldsymbol{r}_2 - \boldsymbol{r}_1$, we can write

$$\boldsymbol{s} = \Delta\boldsymbol{r} = \boldsymbol{r}_2 - \boldsymbol{r}_1. \qquad (2.5)$$

A displacement vector $\boldsymbol{s}$ can be described directly in terms of its magnitude and direction. You might describe a particular displacement by saying 'a distance of 10 m due north', for instance, or you might say that a displacement has 'a magnitude s = 10 m and is directed at $\theta = 90°$ to the x-axis'.

However, it is often more useful to specify the displacement in terms of its components along a specified pair of orthogonal axes. The technique for doing this is very similar to that used for position vectors in the last subsection and is illustrated in Figure 2.11.

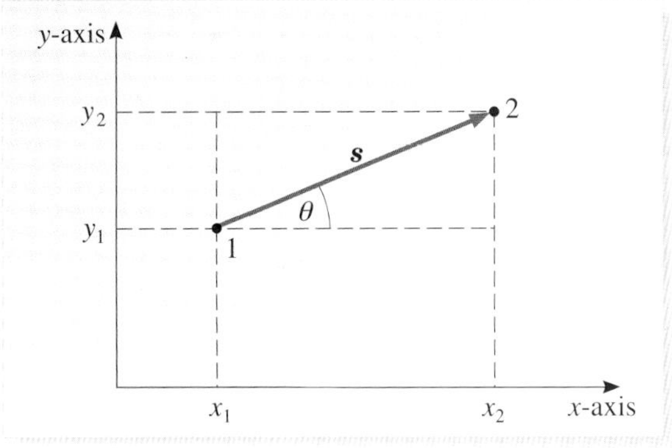

Figure 2.11 The coordinate differences $s_x = \Delta x = x_2 - x_1$ and $s_y = \Delta y = y_2 - y_1$ define the components of the displacement vector $\boldsymbol{s}$.

Once we have chosen particular x- and y-axes, point 1 can be described by coordinate values x_1 and y_1, while point 2 can be described by coordinate values x_2 and y_2. The displacement from point 1 to point 2 will then involve changing coordinates by the amounts $\Delta x = x_2 - x_1$ and $\Delta y = y_2 - y_1$. The quantities Δx and Δy constitute the x- and y-components of the displacement vector and may be given the symbols s_x and s_y.

Thus $\quad s_x = \Delta x = x_2 - x_1 \quad \text{and} \quad s_y = \Delta y = y_2 - y_1$

so $$\boldsymbol{s} = (s_x, s_y) = \Delta \boldsymbol{r} = \boldsymbol{r}_2 - \boldsymbol{r}_1 = (x_2 - x_1, y_2 - y_1). \qquad (2.6)$$

So a displacement vector may be represented by an ordered pair of coordinate differences, just as a position vector can be represented by an ordered pair of coordinates.

In view of the similarities between position vectors and displacement vectors you will not be surprised to learn that similar relationships exist between the magnitudes and directions, and the components in both cases. Thus the magnitude of a displacement vector is related to the displacement's components by

$$s = \sqrt{s_x^2 + s_y^2} \qquad (2.7)$$

and, if the angle between the x-axis and the displacement vector is θ, then

$$s_x = s \cos \theta \qquad (2.8a)$$

$$s_y = s \sin \theta. \qquad (2.8b)$$

Notice that the symbol s representing the magnitude of the displacement is not emboldened.

Any displacement vector $\boldsymbol{s} = (s_x, s_y)$, or $\Delta \boldsymbol{r} = (\Delta x, \Delta y)$ if you prefer, is completely specified by its components. It does not have to start or finish at a particular point, it just has to have a definite magnitude and direction. In this sense, displacement vectors are more general than position vectors which are tied to the origin.

Question 2.5 A particle moves in the xy-plane from point A, with position vector (3 m, 2 m) to point B, with position vector (7 m, −1 m). (a) What are the x- and y-components of the particle's displacement vector? (b) What is the magnitude of the displacement vector? (c) What is the angle between the x-axis and the displacement vector?

Question 2.6 What is another name for the components of the displacement vector of a particle from the origin? ■

Combining displacements: the triangle rule

Figure 2.12 provides another view of the moving ball we considered earlier. The ball starts from the origin, travels to A and then on to C. When the ball is at A, its displacement from the origin is given by the vector $\boldsymbol{a}$. In moving from A to C, the ball undergoes a further displacement $\boldsymbol{b}$. The combined effect of displacements $\boldsymbol{a}$ and $\boldsymbol{b}$ is to transport the ball to point C, where its displacement from the origin is the displacement vector $\mathbf{c}$. We can summarize this by saying that $\mathbf{c}$ is the **resultant** of $\boldsymbol{a}$ and $\boldsymbol{b}$, and by writing

$$\boldsymbol{a} + \boldsymbol{b} = \mathbf{c}. \tag{2.9}$$

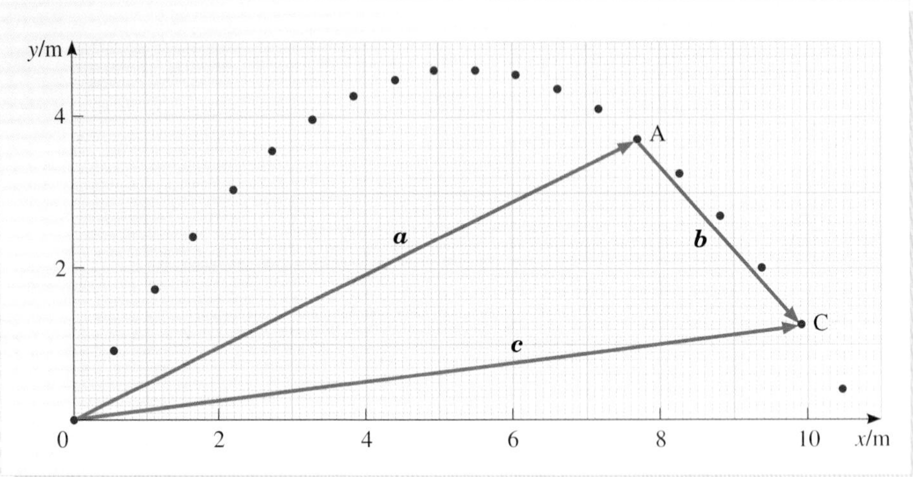

Figure 2.12 The displacement from the origin to point C can be regarded as the combination of a displacement from the origin to A, followed by a displacement from A to C.

Remember, $\boldsymbol{a}$ and $\boldsymbol{b}$ are not simply numbers, they are displacements that involve directions as well as magnitudes, so we have to be very clear about what we mean when we add them together. Figure 2.12 is clear, but it is nonetheless helpful to spell out its message in words:

> Suppose $\boldsymbol{a}$ and $\boldsymbol{b}$ are displacement vectors represented diagrammatically by arrows of appropriate length and direction. If the arrow representing $\boldsymbol{b}$ is drawn from the head of the arrow representing $\boldsymbol{a}$, then an arrow drawn from the tail of $\boldsymbol{a}$ to the head of $\boldsymbol{b}$ represents their resultant, the displacement $\mathbf{c} = \boldsymbol{a} + \boldsymbol{b}$.

This statement is sometimes called the **triangle rule** for adding displacement vectors, since the displacements $\boldsymbol{a}$, $\boldsymbol{b}$ and $\mathbf{c}$ form a triangle, as shown in Figure 2.12.

Knowing the triangle rule, we can easily add displacements diagrammatically, once we have drawn them to an appropriate scale. However, drawing accurate diagrams can be tedious, especially if the displacements are specified in terms of their components rather than in terms of magnitudes and directions. It would therefore be useful to find a mathematical rule relating the components c_x and c_y of $\mathbf{c}$ to the components of $\boldsymbol{a}$ and $\boldsymbol{b}$. You can deduce this rule for yourself by answering the following question.

Question 2.7 Measure the x- and y-components of the three displacements shown in Figure 2.12. Look carefully at your answers and suggest a general rule for relating the components of $\mathbf{c}$ to the components of $\boldsymbol{a}$ and $\boldsymbol{b}$. ■

You should have found from Question 2.7 that the x- and y-components of $\boldsymbol{c}$ are equal to the sums of the corresponding components of $\boldsymbol{a}$ and $\boldsymbol{b}$. That is:

if $\quad \boldsymbol{c} = \boldsymbol{a} + \boldsymbol{b}$

then $\quad c_x = a_x + b_x \quad$ and $\quad c_y = a_y + b_y$.

So, the components of a sum of two displacements are given by the sums of the components of those displacements:

$$\boldsymbol{a} + \boldsymbol{b} = (a_x, a_y) + (b_x, b_y) = (a_x + b_x, a_y + b_y). \qquad (2.10)$$

Note that it is the components of the displacements that add, and *not* their magnitudes. In fact, it is generally the case that

$c \neq a + b$, (where the symbol $\neq$ means 'is not equal to')

and the rule relating c to a and b is rather complicated. This is one reason why it is often useful to employ a component description of displacements.

To end this subsection we consider the operation of multiplying a displacement vector by a number. This is related to the addition of displacement vectors because a quantity such as $2\boldsymbol{a}$ (the result of multiplying the displacement vector $\boldsymbol{a}$ by the number 2) can be interpreted as $\boldsymbol{a} + \boldsymbol{a}$. Since we already know that $\boldsymbol{a} + \boldsymbol{a} = (2a_x, 2a_y)$, we can write

$$2\boldsymbol{a} = \boldsymbol{a} + \boldsymbol{a} = (2a_x, 2a_y).$$

Note that multiplying a displacement vector by 2 has the effect of multiplying *each* of its components by 2. This is shown graphically in Figure 2.13a. The displacement vector $2\boldsymbol{a}$ points in the same direction as $\boldsymbol{a}$, but its magnitude is twice that of $\boldsymbol{a}$. This idea can be generalized by requiring that

$$\lambda\boldsymbol{a} = (\lambda a_x, \lambda a_y) \qquad (2.11)$$

for any number, λ. In other words, the effect of multiplying a displacement vector by a number is to multiply each of the components of the displacement vector by that number, whatever its value.

If λ is positive, the displacement vector $\lambda\boldsymbol{a}$ points in the same direction as $\boldsymbol{a}$, but has magnitude λa. If λ is negative, $\lambda\boldsymbol{a}$ points in the opposite direction to $\boldsymbol{a}$ but its magnitude must still be a positive quantity, so it will be equal to $-\lambda a$ when λ is negative. The case when $\lambda = -2$ is illustrated in Figure 2.13b.

When $\lambda = -1$, $\lambda\boldsymbol{a} = (-1)\boldsymbol{a} = -\boldsymbol{a}$, a vector having the same magnitude as $\boldsymbol{a}$ but pointing in the opposite direction.

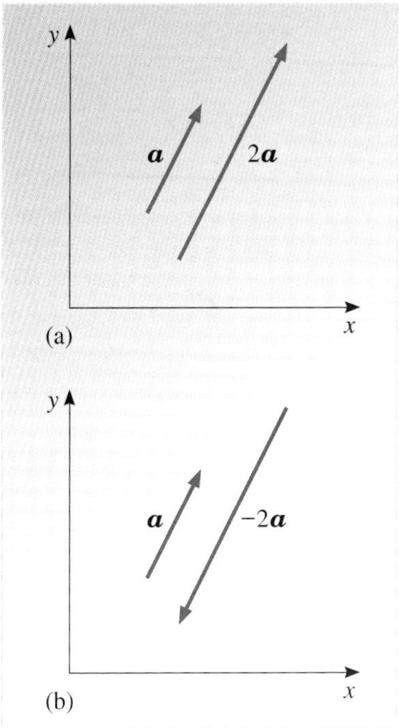

Figure 2.13 (a) The result of multiplying a displacement vector $\boldsymbol{a}$ by 2 is another displacement vector, denoted $2\boldsymbol{a}$, that points in the same direction as $\boldsymbol{a}$ but has twice the magnitude. Note that since displacements are not tied to any particular position, $2\boldsymbol{a}$ is parallel to $\boldsymbol{a}$ but need not lie along the same line. (b) The displacements $\boldsymbol{a}$ and $-2\boldsymbol{a}$.

Question 2.8 Suppose that the displacement $\boldsymbol{c}$ is the resultant of displacements $\boldsymbol{a}$ and $\boldsymbol{b}$. In what circumstances do the corresponding magnitudes add, so that $c = a + b$? [*Hint*: Draw a sketch representing the sum of $\boldsymbol{a}$ and $\boldsymbol{b}$, using the triangle rule.]

Question 2.9 Suppose that $\boldsymbol{a} = (3\,\text{m}, 4\,\text{m})$ and $\boldsymbol{b} = (-1\,\text{m}, 2\,\text{m})$.

(a) What are the components of the displacement $\boldsymbol{a} - 3\boldsymbol{b}$? [Note that the displacement $\boldsymbol{a} - 3\boldsymbol{b}$ means the *addition* of the displacements $\boldsymbol{a}$ and $-3\boldsymbol{b}$, i.e. $\boldsymbol{a} - 3\boldsymbol{b} = \boldsymbol{a} + (-3\boldsymbol{b})$.]

(b) If $\boldsymbol{c} = \boldsymbol{a} + \boldsymbol{b}$, calculate the magnitudes a, b and c. Does $c = a + b$? ■

2.4 A note on vectors and scalars

Quantities that have a definite magnitude and direction are common in physics. Such a quantity is generally referred to as a *vector quantity*, or, more simply, a **vector**. A displacement vector is a good example of such a quantity but there are many others, including velocity, acceleration and force. A quantity such as mass or temperature that has no direction associated with it is called a **scalar**. The value of a scalar quantity can be completely specified as the product of a number and an appropriate unit of measurement (e.g. a mass of 20 kg or a temperature of −4 °C). This subsection lists some of the important properties of vectors and scalars, together with some associated notational conventions. The discussion is restricted to two-dimensional vectors, specifically vectors that lie in the xy-plane and can therefore be specified by their x- and y-components. However, you should note that these results can easily be generalized to three dimensions.

1 First, and most important, a vector has a magnitude and a direction.

2 Vectors are printed using bold letters $\boldsymbol{r}$, $\boldsymbol{s}$, $\boldsymbol{v}$, ..., and their magnitudes are printed using ordinary letters r, s, v, ... In handwritten work, vectors should be indicated by a curly underline $\underset{\sim}{\mathrm{r}}$, $\underset{\sim}{\mathrm{s}}$, $\underset{\sim}{\mathrm{v}}$, ... and magnitudes written normally, with no underlining.

3 Any vector quantity $\boldsymbol{f}$ can be represented diagrammatically by an arrow. The direction of the arrow represents the direction of $\boldsymbol{f}$, and the length of the arrow represents the magnitude of $\boldsymbol{f}$.

4 Given a pair of orthogonal x- and y-axes, a vector $\boldsymbol{f}$, lying in the xy-plane, can be specified by its x- and y-components, f_x and f_y. The vector may then be written

$$\boldsymbol{f} = (f_x, f_y). \qquad (2.12)$$

The components are given by

$$f_x = f \cos\theta \quad \text{and} \quad f_y = f \sin\theta, \qquad (2.13)$$

where f is the magnitude of the vector, and is given by

$$f = \sqrt{f_x^2 + f_y^2} \qquad (2.14)$$

and θ is the angle (measured anticlockwise) from the positive x-axis to the direction of $\boldsymbol{f}$. For $0° \leq \theta \leq 90°$, $\theta = \arcsin(f_y/f)$.

This process of finding components from magnitudes and directions turns out to be very useful and is called **resolution**. So far we have only considered resolution along the x- and y-axes, but the process can be generalized so that you can find the component of a given vector along any given direction.

5 The components f_x and f_y of a vector $\boldsymbol{f}$ are scalar quantities and may be positive, negative or zero, depending on the direction of the vector relative to the coordinate axes. The magnitude f of $\boldsymbol{f}$ is also a scalar, but can never be negative.

An alternative notation for the magnitude of $\boldsymbol{f}$ that emphasizes the fact that it cannot be negative is $|\boldsymbol{f}|$.

6 Equations can be written in terms of vectors, e.g. $\boldsymbol{f} = \boldsymbol{g}$. This equation means that:

(a) The magnitude of $\boldsymbol{f}$ is equal to the magnitude of $\boldsymbol{g}$.

(b) The direction of $\boldsymbol{f}$ is the same as the direction of $\boldsymbol{g}$.

(c) The components of $\boldsymbol{f}$ are equal to the components of $\boldsymbol{g}$:

$$f_x = g_x \quad \text{and} \quad f_y = g_y.$$

Statements (a) and (b) together are equivalent to statement (c).

Conversely, if two vectors have the same magnitudes and directions (and consequently the same components), they are said to be equal. Vectors may only be equated with other vectors. It does not make any sense to write $\boldsymbol{f} = g$, for example. If all the components of a vector are zero, we write $\boldsymbol{f} = \mathbf{0}$, where the bold symbol **0** represents the zero vector, (0, 0).

7 Two vectors of the same type (for example, two displacements or two velocities) can be added together to form a resultant. The resultant of $\boldsymbol{f} = (f_x, f_y)$ and $\boldsymbol{g} = (g_x, g_y)$ is found by adding the corresponding components:

$$\boldsymbol{f} + \boldsymbol{g} = (f_x + g_x, f_y + g_y). \qquad (2.15)$$

The triangle rule provides a geometric interpretation of vector addition. (See Figure 2.14.)

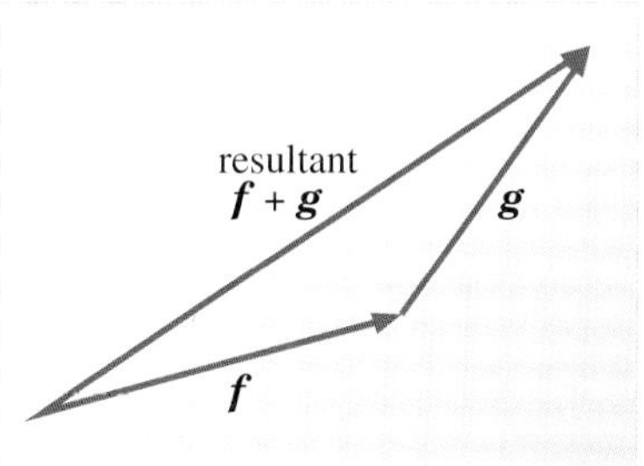

Figure 2.14 Two vectors can be added together graphically, using the triangle rule.

8 Any vector $\boldsymbol{f}$ can be multiplied by a scalar quantity λ (which may be positive or negative and which is not necessarily a whole number). The result is a vector whose components are λ times the components of $\boldsymbol{f}$:

$$\lambda\boldsymbol{f} = (\lambda f_x, \lambda f_y). \qquad (2.16)$$

If λ is a positive quantity, the vector $\lambda\boldsymbol{f}$ points in the same direction as $\boldsymbol{f}$. If λ is a negative quantity, $\lambda\boldsymbol{f}$ points in the opposite direction to $\boldsymbol{f}$.

Question 2.10 Figure 2.15 shows two displacement vectors $\boldsymbol{a}$ and $\boldsymbol{b}$. Both vectors have the same direction and the same magnitude, but they have different initial and final points. (a) Can we say that $\boldsymbol{a} = \boldsymbol{b}$ in this case? (b) Can we say that $|\boldsymbol{a} + \boldsymbol{b}| = a + b$ in this case? (*Reminder*: $|\boldsymbol{a} + \boldsymbol{b}|$ means the magnitude of the vector obtained by adding $\boldsymbol{a}$ and $\boldsymbol{b}$.) ■

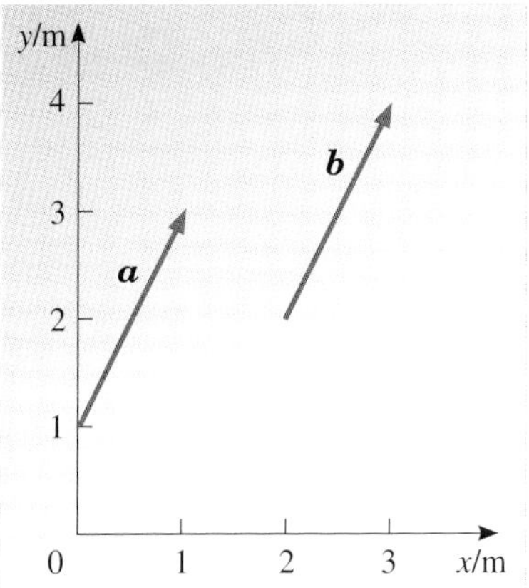

Figure 2.15 Two displacement vectors. (Note that 1 cm on the diagram represents 1 m in real space.)

3 Velocity and acceleration in a plane

3.1 Velocity in a plane

In Chapter 1, we defined velocity as the rate of change of position. Recall that we were concerned with the motion of a particle along a straight line (the x-axis). For a particle moving uniformly from position x_1 at time t_1 to position x_2 at time t_2, the (constant) velocity was defined to be

$$v_x = \frac{\Delta x}{\Delta t} \qquad \text{(Eqn 1.5)}$$

where $\Delta x = x_2 - x_1$ is the change in position that occurs in the time interval $\Delta t = t_2 - t_1$.

For a particle moving non-uniformly, the *instantaneous velocity* at a particular time was given by the *instantaneous* rate of change of position. This could be determined from the gradient of a position–time graph and was written as a derivative:

$$v_x = \frac{dx}{dt}. \qquad \text{(Eqn 1.14)}$$

Given x as a *function* of t, you also learned how to work out the derivative algebraically in a number of simple cases.

Similar ideas apply in two dimensions, but now position is defined by a position *vector*, and so it would seem reasonable to define the velocity *vector* as the rate of change of the position vector. To see what is involved, have a look at Figure 2.16, which returns to the example of the thrown ball. This diagram is the same as Figure 2.12, but uses a different notation in order to emphasize the position vector and its changes.

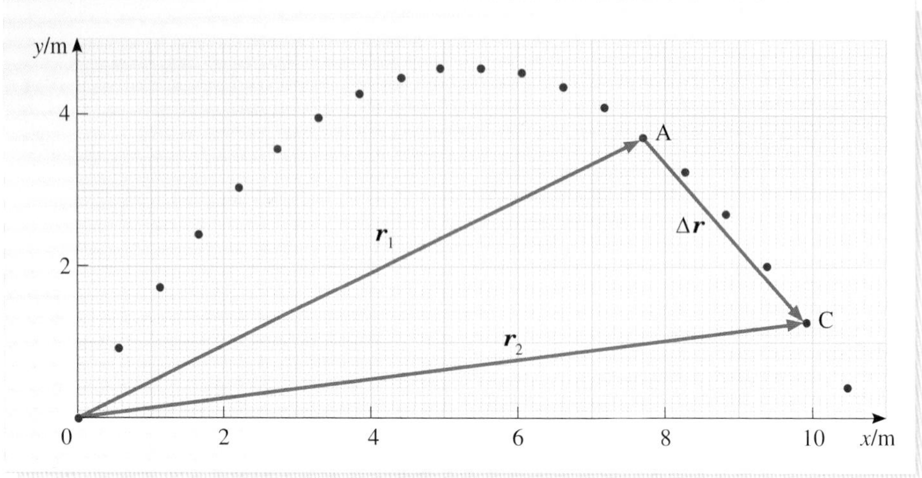

Figure 2.16 The average velocity vector of the ball over the interval from t_1 to t_2 is found by taking the ratio $\Delta \boldsymbol{r}/\Delta t$. This is the average rate of change of the particle's position vector divided by the time interval Δt.

At time t_1, the ball is at position A and its position vector is $\boldsymbol{r}_1$. At a later time t_2, the ball is at position C and its position vector is $\boldsymbol{r}_2$. The displacement from A to C is $\Delta \boldsymbol{r}$. This displacement vector represents the change in position of the ball in the time interval $\Delta t = t_2 - t_1$. The **average velocity vector** in this time interval is defined to be

$$\langle \boldsymbol{v} \rangle = \frac{\Delta \boldsymbol{r}}{\Delta t}. \tag{2.17}$$

The average velocity $\langle \boldsymbol{v} \rangle$ is a vector quantity that points in the same direction as the displacement vector $\Delta \boldsymbol{r}$ since Δt is always positive. Like all vector quantities, $\langle \boldsymbol{v} \rangle$ can be described in terms of its components. Now, $\Delta \boldsymbol{r}$ has components Δx and Δy and can be written as

$$\Delta \boldsymbol{r} = (\Delta x, \Delta y).$$

The average velocity vector is found by dividing both sides by Δt or, equivalently, multiplying both sides by $1/\Delta t$. In this way we obtain

$$\langle \boldsymbol{v} \rangle = \frac{\Delta \boldsymbol{r}}{\Delta t} = \left(\frac{\Delta x}{\Delta t}, \frac{\Delta y}{\Delta t} \right). \tag{2.18}$$

Thus, each component of the average velocity vector is found by dividing the change in the corresponding coordinate by the change in time.

Question 2.11 If the ball in Figure 2.16 is at point A at $t = 1.4$ s and at point C at $t = 1.8$ s, what is the average velocity vector of the ball over the time interval from $t = 1.4$ s to $t = 1.8$ s? (*Hint*: Look at your answer to Question 2.7.) ■

Of course, the velocity of the moving ball is changing throughout the motion, so what we really want to know is the **instantaneous velocity vector $\boldsymbol{v}$** of the ball. In view of the expression for the average velocity you will not be surprised to learn that the instantaneous velocity vector is given by

$$\boldsymbol{v} = \frac{d\boldsymbol{r}}{dt} = \left(\frac{dx}{dt}, \frac{dy}{dt}\right). \tag{2.19}$$

For brevity, the word 'instantaneous' is usually omitted and 'velocity' is generally taken to mean the 'instantaneous velocity at a given time'.

Equation 2.19 shows that the velocity vector is the derivative of the position vector, and this derivative is found by regarding each of the particle's position coordinates as a function of time and differentiating each coordinate separately. The components of the velocity vector are

$$v_x = \frac{dx}{dt} \quad \text{and} \quad v_y = \frac{dy}{dt}. \tag{2.20}$$

The x-component of the particle's velocity is the rate of change of its x-coordinate with respect to time, and the y-component is the rate of change of the y-coordinate with respect to time.

Question 2.12 Suppose that throughout its flight, the coordinates of the ball in Figure 2.3 depend on time according to the equations $x = At$ and $y = Bt - Ct^2$. If $A = 5.5\,\text{m s}^{-1}$, $B = 9.5\,\text{m s}^{-1}$ and $C = 4.9\,\text{m s}^{-2}$, use the standard derivatives in Table 1.6 to calculate the components of the ball's velocity at $t = 0.6\,\text{s}$. ■

Like all vectors, the velocity vector has both magnitude and direction. The magnitude of the velocity is known as the (instantaneous) **speed** of the particle and the direction of the velocity is the (instantaneous) direction of motion of the particle. Like all vectors, the velocity vector can be represented on a diagram by drawing an arrow. The arrow in Figure 2.17 represents the velocity of the ball at time $t = 0.6\,\text{s}$. Notice that the scales along the horizontal and vertical axes are now calibrated in m s^{-1}, in order to represent velocities. The origin of *these* axes corresponds to the particle being at rest, and *not* to the particle being at the point with coordinates $x = y = 0$. The x- and y-components of the velocity are found by constructing perpendiculars to the v_x and v_y axes. You can confirm that this gives $v_x = 5.5\,\text{m s}^{-1}$ and $v_y = 3.6\,\text{m s}^{-1}$, in agreement with Question 2.12.

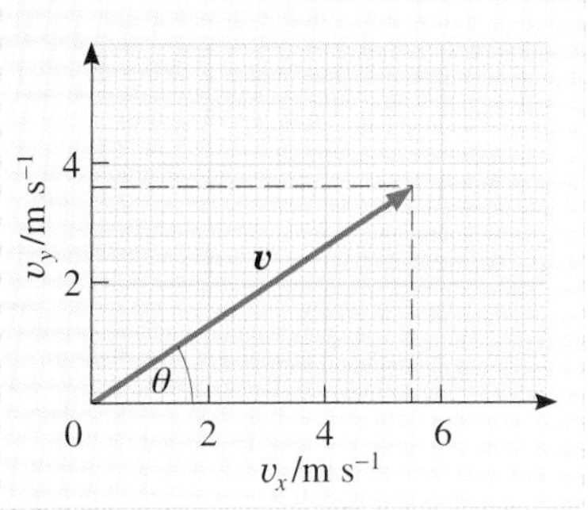

Figure 2.17 The velocity vector of the ball at 0.6 s after launch.

The speed v of the ball (at time $t = 0.6\,\text{s}$) is represented by the length of the arrow in Figure 2.17. This can be measured approximately with a ruler and converted into a speed by applying the appropriate scale factor. Alternatively, Pythagoras' theorem can be used to obtain the speed from the components, using

$$v = \sqrt{v_x^2 + v_y^2}\,.$$

Conversely, the velocity components v_x and v_y can be expressed in terms of the speed v and the angle θ between the v_x axis and the velocity vector. Applying Equation 2.13 to the vector of Figure 2.17, gives

$$v_x = v\cos\theta \quad \text{and} \quad v_y = v\sin\theta.$$

Question 2.13 (a) Use your answer to Question 2.12 to find the speed of the ball at $t = 0.6$ s. (b) In what direction is the ball in Question 2.12 travelling at $t = 0.6$ s? Express your answer by giving the angle θ between the horizontal x-axis and the ball's direction of motion.

Question 2.14 Suppose the position of a particle at time t is given by $x = At - Bt^2$ and $y = At - Bt^2$, where $A = 9.5\ \mathrm{m\,s^{-1}}$ and $B = 4.9\ \mathrm{m\,s^{-2}}$. Find the general expressions for the velocity components v_x and v_y of this particle and evaluate these components when $t = 0$. ■

3.2 Acceleration in a plane

In Chapter 1 we defined (instantaneous) acceleration as the rate of change of velocity, but restricted our attention to one dimension (the x-axis). The acceleration along the x-axis was given by the derivative of the velocity v_x with respect to time:

$$a_x = \frac{\mathrm{d}v_x}{\mathrm{d}t} = \frac{\mathrm{d}^2 x}{\mathrm{d}t^2}.$$

In two dimensions acceleration is described by a vector and is defined to be the rate of change of velocity. Thus the (instantaneous) **acceleration vector $\boldsymbol{a}$** is the derivative of the velocity vector and its components are the derivatives of the velocity components:

$$\boldsymbol{a} = \frac{\mathrm{d}\boldsymbol{v}}{\mathrm{d}t} = \left(\frac{\mathrm{d}v_x}{\mathrm{d}t}, \frac{\mathrm{d}v_y}{\mathrm{d}t}\right). \tag{2.21}$$

The components of the acceleration vector are therefore

$$a_x = \frac{\mathrm{d}v_x}{\mathrm{d}t} \quad \text{and} \quad a_y = \frac{\mathrm{d}v_y}{\mathrm{d}t}. \tag{2.22}$$

Remember that a derivative tells us the rate at which one quantity changes in response to changes in another quantity. The components of the acceleration vector therefore tell us the rate at which the velocity components change in time.

Like any vector, the acceleration vector has a magnitude, a, related to its components by using Equation 2.14

$$a = \sqrt{a_x^2 + a_y^2}.$$

Conversely, the acceleration components are found by using Equation 2.13

$$a_x = a\cos\theta \quad \text{and} \quad a_y = a\sin\theta$$

where θ is the angle between the x-axis and the acceleration vector.

Question 2.15 Using the equations given in Question 2.12, calculate: (a) the acceleration vector of the ball in Figure 2.3 at time t; (b) the magnitude of the acceleration; (c) the direction of the acceleration. ■

The answer to Question 2.15 shows that in this case the acceleration vector is a constant, points vertically downwards, and has a magnitude of about $9.8\ \mathrm{m\,s^{-2}}$. This is consistent with our discussion of terrestrial gravity in Chapter 1, where we saw that gravity acts vertically downwards and produces a downward acceleration of magnitude $9.8\ \mathrm{m\,s^{-2}}$.

Some students have difficulty visualizing what is meant by a constant downward acceleration of $9.8\ \mathrm{m\,s^{-2}}$ in a situation like that shown in Figure 2.3, where a particle travels along a curved path. The meaning is simply that, in every second, the magnitude of the change in the velocity vector is $9.8\ \mathrm{m\,s^{-1}}$, and this change in velocity is directed vertically downwards. For example, Figure 2.18 shows two velocity vectors $\boldsymbol{v}_1$ and $\boldsymbol{v}_2$ corresponding to times $t_1 = 0.6\ \mathrm{s}$ and $t_2 = 0.8\ \mathrm{s}$. The fact that these two velocity vectors are different means that the ball is accelerating. In Figure 2.19, the two velocities $\boldsymbol{v}_1$ and $\boldsymbol{v}_2$ are drawn as arrows with their two tails touching. This allows us to show the change in velocity, which we have labelled $\Delta\boldsymbol{v}$. It is clear from the triangle rule that $\Delta\boldsymbol{v}$ points vertically downwards and has magnitude $2.0\ \mathrm{m\,s^{-1}}$. This corresponds to a constant downward acceleration of magnitude $2.0\ \mathrm{m\,s^{-1}}/(0.8 - 0.6)\ \mathrm{s} = 10\ \mathrm{m\,s^{-2}}$, which is consistent with $9.8\ \mathrm{m\,s^{-2}}$ to within the accuracy of the diagram.

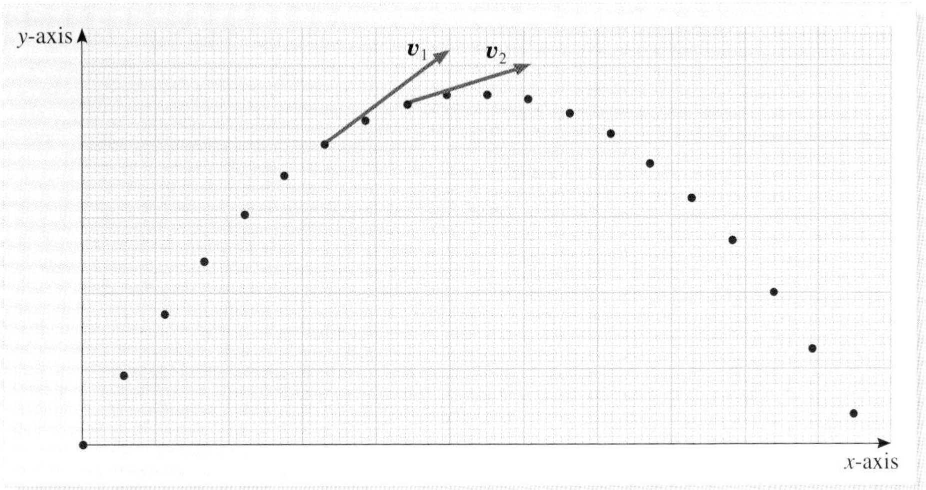

Figure 2.18 Velocity vectors $\boldsymbol{v}_1$ and $\boldsymbol{v}_2$ for the ball at times $t_1 = 0.6\ \mathrm{s}$ and $t_2 = 0.8\ \mathrm{s}$.

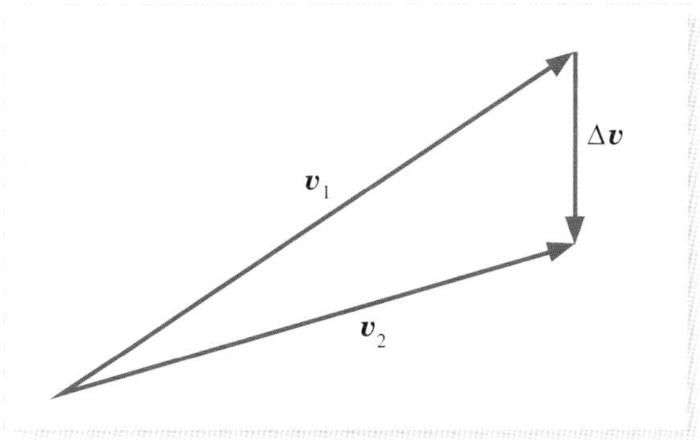

Figure 2.19 The initial velocity $\boldsymbol{v}_1$, the final velocity $\boldsymbol{v}_2$, and the change in velocity $\Delta\boldsymbol{v}$ are related by the triangle rule. This diagram is drawn to a scale of 1 cm to $1\ \mathrm{m\,s^{-1}}$, so it is apparent that the change in velocity points vertically downwards and has a magnitude of about $2.0\ \mathrm{m\,s^{-1}}$.

Note that the constant downward acceleration does not prevent the particle from moving upwards. It simply means that an upward velocity would decrease at a steady rate. (Similarly, any downward velocity would increase at a steady rate due to a constant downward acceleration.)

One final point can be made from Figure 2.19. Notice that the arrows representing $\boldsymbol{v}_1$ and $\boldsymbol{v}_2$ differ only slightly in length (by much less than the length of the arrow for $\Delta\boldsymbol{v}$). This means that the *speed* of the ball changes at a *slower* rate than $9.8\ \mathrm{m\,s^{-2}}$. If this seems puzzling, you should realize that the acceleration of the ball is partly due to a change in speed and partly due to a change in the direction of motion. In Chapter 3, you will meet the case of uniform circular motion, in which the speed of the particle remains fixed but acceleration occurs because the direction of motion continuously changes.

Question 2.16 At a certain time a particle undergoing constant acceleration has velocity $\boldsymbol{v}_1 = (10, 0)\,\mathrm{m\,s^{-1}}$. One second later, the particle's velocity is $\boldsymbol{v}_2 = (0, 10)\,\mathrm{m\,s^{-1}}$. Find the following for this one-second interval:

(a) the change in velocity;

(b) the acceleration;

(c) the magnitude of the acceleration;

(d) the direction of the acceleration;

(e) the change in the speed, Δv;

(f) the magnitude of the change in the velocity, $|\Delta \boldsymbol{v}|$.

What is the physical difference between Δv and $|\Delta \boldsymbol{v}|$? ■

4 Projectile motion

4.1 Introducing projectile motion

The term **projectile** is used to describe any object that is launched into the air near the Earth's surface and which thereafter moves in unpowered flight in such a way that its motion is determined by the effects of gravity and air resistance. Tennis balls, golf balls and footballs all provide good examples of projectiles, as do javelins, high jumpers and long jumpers. Some examples of projectiles being launched are shown in Figure 2.20.

Figure 2.20 Some examples of projectiles being launched.

In what follows we shall ignore air resistance and model all projectiles as particles; even so, many of the results we obtain will be approximately correct and will certainly be of considerable interest.

The study of projectiles has long been considered an important problem in science. In medieval times, the paths of cannon-balls and arrows were of interest to the military, so there were very practical reasons for seeking to understand the way in which projectiles moved. The first person to achieve such an understanding was the Italian scientist Galileo Galilei.

The essential point which others had missed but which Galileo grasped was this:

> The horizontal and the vertical movements of a projectile are independent, apart from the fact that they must both have the same duration.

In particular in the absence of air friction:

1 The vertical component of a projectile's motion is an example of *uniformly accelerated motion* in which the constant acceleration caused by gravity is directed downward and has magnitude g (with the approximate value $9.8\,\mathrm{m\,s^{-2}}$).

2 The horizontal component of a projectile's motion is an example of *uniform motion* in which the constant horizontal velocity is determined when the projectile is launched.

So a projectile travels with constant velocity in the horizontal direction, during the time that it takes to rise and fall under constant downward acceleration in the vertical direction.

By combining this principle with information about the initial position and velocity, Galileo was able to predict the motion of projectiles such as cannon-balls. Galileo's many scientific achievements, of which this may be the greatest, have caused a number of commentators to regard him as the first true physicist in the modern sense.

Galileo Galilei (1564–1642)

Galileo Galilei (Figure 2.21), the son of a musician, was born in Pisa, Italy, in 1564. He embarked on the study of medicine at the University of Pisa, but became interested in mathematics and soon thereafter discovered that a pendulum executing small oscillations would complete a full swing in a time that was independent of the extent of the swing, as long as it remained small. He gave up his medical studies and spent the next four years in nearby Florence, studying a variety of subjects, before returning to Pisa with a temporary appointment as a professor of mathematics in the university.

In 1592 Galileo moved to the University of Padua, which served the powerful independent republic of Venice. His stay in Padua seems to have been mainly happy and productive, even though Galileo suffered from a lack of funds for much of the time. In Padua, Galileo carried out some of his most celebrated studies including those on the motion of accelerated bodies.

The year 1609 marked a turning point in Galileo's life. Having heard of the invention of the telescope, he constructed one for himself and showed it to a number of friends, pointing out its practical value, especially to those at sea. He brought the new invention to the attention of the authorities in Venice (a major maritime power at that time) who rewarded Galileo with a substantial increase in salary. Although Galileo did not invent the telescope, he was the first to report the result of using it for the systematic study of the heavens. In a hurriedly written book called *Siderius Nuncius* (*The Starry Messenger*), published in 1610, he described his discovery of four moons orbiting Jupiter, of the Milky Way's stellar composition, of mountains on the Moon and of spots on the Sun. He also saw the rings of Saturn, though he did not recognize their true nature. These astronomical discoveries made Galileo internationally famous.

Figure 2.21 Galileo Galilei.

In September 1610, Galileo left Padua in order to return to Tuscany where he became chief mathematician to the Grand Duke Cosimo in Florence. Sadly for Galileo, the domain of the Medici was more strongly influenced by the Catholic Church than was the Republic of Venice, and Galileo soon came under attack for his support of the Copernican doctrine that the Earth moved round the Sun. Summoned to Rome in 1616, he was warned to desist from spreading Copernican views, which he did, at least for a while. In 1630, following the appointment of a new Pope, whom he believed to be sympathetic to his views, Galileo published his *Dialogue Concerning the Two Chief Systems of the World* in which he defended Copernicanism (Figure 2.22). As a result, in 1633, he was condemned by the Inquisition and required to renounce his views.

Following his condemnation Galileo returned to Tuscany, where he spent the rest of his life, mostly in a villa at Arcetri, near Florence. Although essentially under house arrest, Galileo continued to pursue his scientific interests and in 1638 published what is widely regarded as his scientific masterpiece, the *Discourses Concerning Two New Sciences*. In this work Galileo deals first with the strength of materials and then moves on to the subject of motion, and its mathematical description. He analyses uniform (constant velocity) motion and describes uniformly accelerated motion, including the motion of bodies moving down inclined planes. (Such bodies are only accelerated by the component of the acceleration due to gravity that is parallel to the plane, so their acceleration is less than that of a falling body and is therefore easier to study.) He then combines uniform horizontal motion with uniformly accelerated motion to determine the motion of projectiles. As Galileo himself says:

'It has been observed that missiles, that is to say projectiles, follow some kind of curved path, but that it is a parabola no one has shown. I will show that it is, together with other things, neither few in number nor less worth knowing, and what I hold to be even more important, they open the door to a vast and crucial science of which these researches constitute the elements; other geniuses more acute than mine will penetrate its hidden recesses.'

Figure 2.22 The frontispiece of Galileo's book *Dialogue Concerning the Two Chief Systems of the World*.

4.2 The equations of projectile motion

The main principles of projectile motion can be illustrated by returning to the example considered earlier of a ball that is thrown into the air at an angle of $\theta = 60^\circ$ to the horizontal. Suppose the ball is launched with an initial speed $u = 11.0\,\text{m}\,\text{s}^{-1}$ and subsequently moves along a curved path in a vertical plane. We can use Galileo's ideas to predict the time of flight of the ball, the horizontal range of the throw and the shape of the ball's path. In the course of making these predictions, we will develop general equations that are useful in many problems involving projectiles.

For simplicity, we shall start by supposing that the ball is thrown from ground level. Figure 2.23 shows a suitable choice of coordinate system — the x-axis is horizontal, the y-axis points vertically upwards and the ball moves in the xy-plane. The initial velocity vector then has components

$$u_x = u \cos \theta \qquad (2.23a)$$

$$u_y = u \sin \theta. \qquad (2.23b)$$

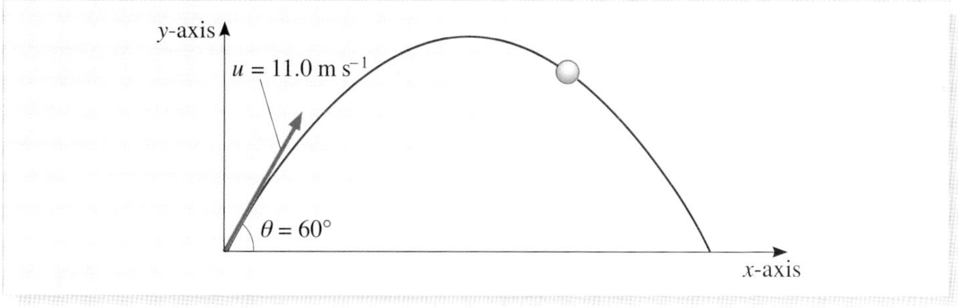

Figure 2.23 A ball is thrown at a given speed and angle, from ground level. How long will the ball spend in the air? How far will it travel? What will be the shape of its path?

Using the known values, $u = 11.0\,\mathrm{m\,s^{-1}}$ and $\theta = 60°$, we obtain

$$u_x = 5.5\,\mathrm{m\,s^{-1}} \quad \text{and} \quad u_y = 9.5\,\mathrm{m\,s^{-1}}. \qquad (2.24)$$

Now, ignoring the effects of air resistance, the components of the projectile's acceleration are

$$a_x = 0 \qquad (2.25a)$$

$$a_y = -g. \qquad (2.25b)$$

Notice that g is a positive quantity, so a_y is negative.

Using these values and Galileo's observation that the horizontal and vertical movements of a projectile are independent, we can now set about describing the x- and y-components of the motion.

Motion in the x-direction

The x-component of the acceleration is zero, so the x-component of velocity remains constant; that is

$$v_x = u_x. \qquad (2.26)$$

Consequently, the x-component of the displacement (from the projectile's initial position) is proportional to time, so

$$s_x = u_x t. \qquad (2.27)$$

Equations 2.26 and 2.27 are the uniform motion equations that we obtained in Chapter 1.

Motion in the y-direction

The y-component of the acceleration has a constant value ($a_y = -g$), so the y-component of velocity is a linear function of time

$$v_y = u_y + a_y t. \qquad (2.28)$$

(This ensures $\dfrac{dv_y}{dt} = a_y$.)

It follows that the y-component of the displacement (from the projectile's initial position) will then be a quadratic function of time

$$s_y = u_y t + \tfrac{1}{2} a_y t^2. \qquad (2.29)$$

(This ensures $\dfrac{ds_y}{dt} = v_y$.)

Equations 2.28 and 2.29 are two of the constant acceleration equations of Chapter 1, with the subscript x replaced by y because we are now dealing with the y-component of the ball's motion. Using the fact that $a_y = -g$ in this case, gives

$$v_y = u_y - gt \qquad (2.30)$$

and

$$s_y = u_y t - \tfrac{1}{2} g t^2. \qquad (2.31)$$

In the next three subsections we use Equations 2.26, 2.27, 2.30 and 2.31 to predict the time of flight, the horizontal range and the path of a projectile.

4.3 The time of flight of a projectile

The time that a projectile spends travelling through the air is called its **time of flight**. What is the time of flight of a projectile launched with speed u at an angle θ to the horizontal? We can answer this question by considering just the vertical motion, using Equation 2.31. At both the beginning and the end of the motion the projectile's vertical component of displacement from its initial position is $s_y = 0$. According to Equation 2.31, the times at which $s_y = 0$ are given by

$$u_y t - \tfrac{1}{2} g t^2 = 0. \qquad (2.32)$$

This can be rewritten by extracting a common factor of t from both terms on the left-hand side, giving

$$t(u_y - \tfrac{1}{2} g t) = 0. \qquad (2.33)$$

Equation 2.33 has two solutions. One is simply $t = 0$ and corresponds to the moment of launch. The other corresponds to the moment when the projectile returns to the ground, and occurs when

$$u_y - \tfrac{1}{2} g t = 0$$

i.e. $t = 2u_y/g$.

This represents the time of flight of the projectile. Thus we can say that for a projectile launched over horizontal ground with initial upward velocity component $u_y = u \sin\theta$, the time of flight T is

$$T = \frac{2u\sin\theta}{g}. \qquad (2.34)$$

Question 2.17 What is the time of flight of the thrown ball we discussed earlier, with $u = 11.0\ \mathrm{m\,s^{-1}}$ and $\theta = 60°$? ■

This is a good place to pause and look at Equation 2.34 and think about what it means i.e. try to read it, as you would a sentence. It says that the time of flight depends on the launch speed, the angle of projection and the magnitude of the acceleration due to gravity. Are these the factors that you would expect to influence the time of flight? Are they the *only* factors you would expect to be involved? If there are any others, why are they absent from Equation 2.34? Is their absence a surprise, or merely a consequence of the approximations we have made? Equation 2.34 indicates that the time of flight will increase if either u or $\sin\theta$ increases, but will decrease if g is increased. Is this the behaviour that you would expect physically? In short, is Equation 2.34 the sort of equation that reveals an astonishing new aspect of the Universe's behaviour, or is it just a compact way of saying that things are pretty much as you would expect them to be? These are the sorts of questions you should ask yourself each time you are presented with a (supposedly) physically meaningful result. By doing so consistently you will gradually deepen your understanding of physics and you will find it easier to spot equations that are misleading or just plain wrong.

4.4 The range of a projectile

The **range** of a projectile is the horizontal distance that it travels during its flight. This distance will be determined by the (constant) horizontal component of the projectile's velocity and its time of flight. Since the horizontal component of velocity is $u_x = u\cos\theta$ and the time of flight is $T = (2u\sin\theta)/g$, it follows from Equation 2.27 that when the projectile returns to the ground, the horizontal component of its displacement will be

$$s_x = u_x T = \frac{2u^2\sin\theta\cos\theta}{g}. \qquad (2.35)$$

This displacement may be positive or negative (it will be positive if θ is between 0° and 90°, but negative if θ is between 90° and 180°). However, its magnitude $|s_x|$ will always be a positive quantity, and it is this that will represent the range R of the projectile. Thus, for a projectile launched at angle θ over horizontal ground with initial speed u, the range is

$$R = \left|\frac{2u^2\sin\theta\cos\theta}{g}\right| \qquad (2.36)$$

Equation 2.36 shows that the range depends strongly on the launch speed u since R depends on the square of u. It also shows that the maximum range, for a given launch speed, is achieved when $\cos\theta\sin\theta$ attains its largest value. If the launch is at too steep an angle, as in Figure 2.24a, then the projectile will not go very far. (For the extreme case of $\theta = 90°$, we would get $R = 0$ since the projectile would go straight up and down.) Neither would the projectile travel very far if launched at too shallow an angle, as in Figure 2.24b. (The extreme case of $\theta = 0°$ would result in the projectile instantly hitting the ground.) It is clear from Figure 2.25 that the maximum value of $\cos\theta\sin\theta$ is 1/2, and that this occurs when $\theta = 45°$. This means that:

For a fixed launch speed u, the maximum possible range R_{max} is

$$R_{max} = u^2/g, \qquad (2.37)$$

and is achieved when the launch angle is 45°.

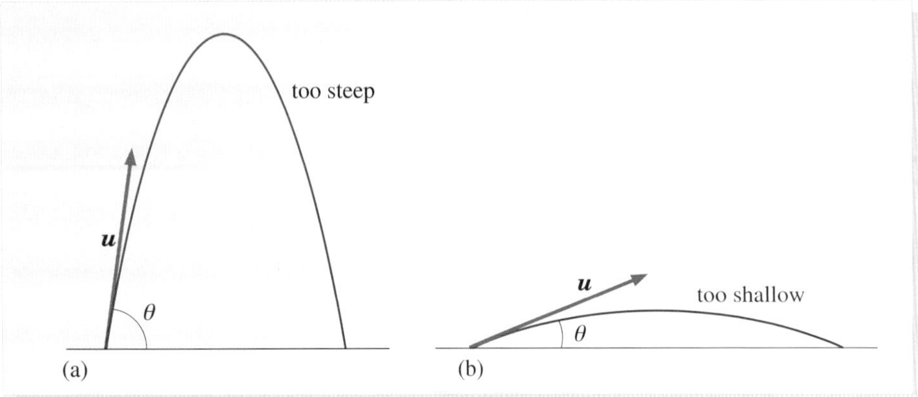

Figure 2.24 Maximizing the range of a projectile.

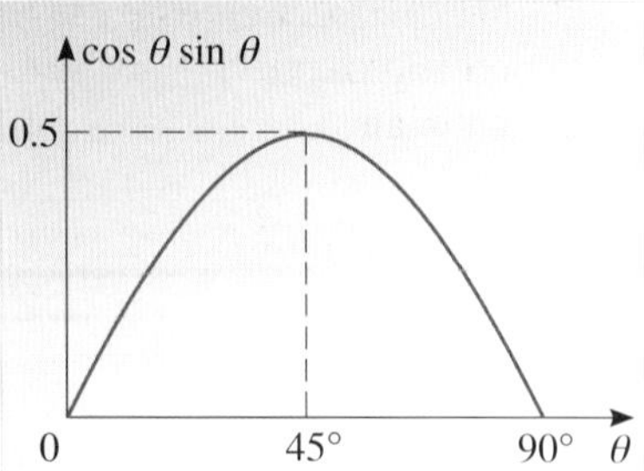

Figure 2.25 A graph of cos θ sin θ against θ for angles between 0° and 90°.

Question 2.18 What is the range of the thrown ball we discussed earlier, with $u = 11.0\,\mathrm{m\,s^{-1}}$ and $\theta = 60°$? What is the range that would have been achieved if the ball had been thrown with the same speed but at 45° to the horizontal? ■

4.5 The trajectory of a projectile

The path followed by a projectile is known as its **trajectory**. One way of specifying the trajectory of a projectile is by saying how the vertical displacement s_y depends on the horizontal displacement s_x. This information can be extracted by eliminating t from Equations 2.27 and 2.31. A simple rearrangement of Equation 2.27 gives $t = s_x/u_x$, and by substituting this value into Equation 2.31 we obtain

$$s_y = u_y\left(\frac{s_x}{u_x}\right) - \frac{1}{2}g\left(\frac{s_x}{u_x}\right)^2. \qquad (2.38)$$

The significance of this can be clarified if we note that for any particular projectile the values of u_x, u_y and g are fixed, so we may rewrite Equation 2.38 in the form

$$s_y = as_x^2 + bs_x \qquad (2.39)$$

where a and b are constants (for a particular projectile) given by

$$a = \frac{-g}{2u_x^2} \quad \text{and} \quad b = \frac{u_y}{u_x}.$$

Equation 2.39 shows that s_y is a *quadratic function* of s_x. (The most general quadratic function would be of the form $as_x^2 + bs_x + c$, but in this case c happens to be 0.) Any curve where one coordinate is a quadratic function of the other coordinate is called a **parabola**. Consequently, we know from Equation 2.39 that:

> A projectile travelling close to the Earth (so that g remains constant), and in the absence of air resistance (so that the acceleration is always directed vertically downwards), follows a trajectory that has the form of a parabola.

Parabolas were well known to the ancient Greeks as members of a wider class of curves called **conic sections**. The origin of this term is shown in Figure 2.26. Galileo must have been delighted to be able to demonstrate that the path of a projectile was, to a good approximation at least, a curve that was already well known to mathematicians.

Question 2.19 Write down the equation of the parabola followed by the thrown ball with $u = 11.0\,\mathrm{m\,s^{-1}}$ and $\theta = 60°$. ■

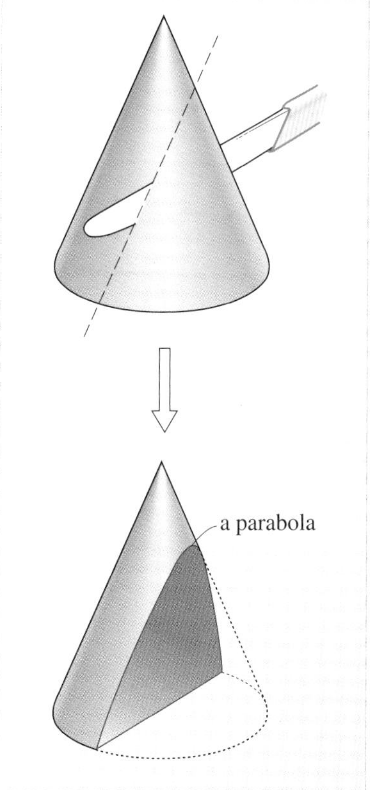

Figure 2.26 The ancient Greeks knew about parabolas in the context of sections through cones. If you stand a cone on its base and make a slice parallel to one edge of the cone, the profile of the cut surface is a parabola.

Box 2.2 Advising a long jumper

So, having studied projectile motion, what should you tell a long jumper? At first sight, the message would seem to be that he or she should try to launch themselves into the air at 45°. However, a moment's reflection should warn you that this is too simple. The 45° result only applies to situations where the take-off speed is fixed. A long jumper leaving the ground at 45° is not likely to attain the same speed as a jumper leaving the ground at a shallower angle, and might therefore suffer a loss of performance. In fact, champion jumpers usually take off at around 20°. Why is this?

A world class long jumper arrives at the take-off board with a speed of about $13\,\mathrm{m\,s^{-1}}$. They attain this by sprinting 30 m or so. As they run this distance they can gradually accelerate, pushing off the track again and again, each time adding to their horizontal velocity. However to take off at 45° they would need the magnitude of their vertical velocity to match that of the horizontal. The vertical component of their velocity can only be gained from one short push during the final foot strike on the take-off board. Because the athlete is running so fast, the foot has very little time in contact with the board. Of course, the athlete modifies the run so as to increase this time as much as possible but they are still limited by the maximum upthrust they can achieve during this time.

The record books bear testimony to the fact that the fastest sprinters are able to jump further than anyone else, if they can master the technique. Many 100 m record holders have also won at the long jump. Jonathan Edwards, the record-breaking triple jumper has been cited as the world's fastest man over 30 m. Clearly the massive horizontal velocity of these athletes enables them to travel a long way. Even though their vertical velocity is limited and they have little time in the air, during this time they are able to travel a great distance. Their technique helps. They stay in contact with the board for as long as possible, pushing vertically upwards; this causes a torque which rotates the body in the air (hence the arm-circling and extended body position to try to minimize rotation, see Figure 2.1). They also take off with a high centre of gravity and land with it as low as possible; this increases the time in the air, enabling them to travel further.

But there is still the tantalizing fact that *if* the take-off angle could be increased, the long jumper would go further. They are faced with a dilemma: when their technique and strength are as good as they can manage, they can only increase their vertical velocity by slowing down. High jumpers take off at about 45°, but they don't go very far. The question for the adviser to answer is *can the technique be refined any further, or is the take-off already at an optimum*?

4.6 A note on quadratic equations

In order to study projectiles further, we must make a diversion into the realm of mathematics and learn how to solve a **quadratic equation**. The general form of a quadratic equation is

$$ax^2 + bx + c = 0 \qquad (2.40)$$

where a, b, c are known constants (these are referred to as *coefficients* and may be numbers, or physical quantities that include units) and a is non-zero. Note that the unknown quantity in Equation 2.40 is x. This unknown quantity should not be confused with the subscript x in Equation 2.39. Nor should the constant a in Equation 2.40 be confused with any of the acceleration components or magnitudes we have been discussing. For the moment, just regard Equation 2.40 as a purely abstract mathematical relation involving three known constants and an unknown quantity x. Here are some specific examples:

(a) $x^2 + 2x = 0$ (i.e. $a = 1$, $b = 2$, $c = 0$)

(b) $2x^2 + 5x + 3 = 0$ (i.e. $a = 2$, $b = 5$, $c = 3$)

(c) $x^2 - 5x + 6 = 0$ (i.e. $a = 1$, $b = -5$, $c = 6$).

Notice that each quadratic equation is obtained by setting a quadratic expression equal to zero. There is only one unknown in each equation (which we have chosen to call x), and when we say that we want to find the **solution** of a quadratic equation we mean that given the values of a, b and c we want to find all those values of x that will cause the equation to be true. Quadratic equations generally have two solutions, i.e. two values of x that make the equation true. For the three equations given above you can confirm (by substituting each of the given values into the equation) that the solutions are:

(a) $x = 0$ and $x = -2$

(b) $x = -1.5$ and $x = -1$

(c) $x = 3$ and $x = 2$.

There are essentially two ways to solve a quadratic equation.

Method 1: Solution by formula

Given a general quadratic equation

$$ax^2 + bx + c = 0 \qquad \text{(Eqn 2.40)}$$

its two solutions are always given by the **quadratic equation formula**

$$x = \frac{-b \pm \sqrt{b^2 - 4ac}}{2a}. \qquad (2.41)$$

Note the symbol ± that occurs in this formula. This is read as 'plus or minus' and is an instruction to take both possibilities into account. Equation 2.41 thus describes two solutions to Equation 2.40

$$x = \frac{-b + \sqrt{b^2 - 4ac}}{2a} \quad \text{and} \quad x = \frac{-b - \sqrt{b^2 - 4ac}}{2a}.$$

It is worth noting that any particular quadratic equation will satisfy one of the following conditions:

- If $b^2 > 4ac$ (i.e. if b^2 is *greater than* $4ac$) then the formula straightforwardly yields two distinct solutions;
- If $b^2 = 4ac$ then the two solutions become identical;
- If $b^2 < 4ac$ (i.e. if b^2 is *less than* $4ac$) then each of the two solutions involves the square root of a negative number and is represented by what is referred to as a *complex number*. You will not be required to deal with such cases in this book.

Example 2.1

Solve the equation $2x^2 + 5x + 3 = 0$ for x, and check the correctness of your solutions.

Solution

Setting $a = 2$, $b = 5$ and $c = 3$ in the quadratic equation formula, gives

$$x = \frac{-5 \pm \sqrt{5^2 - 4 \times 2 \times 3}}{2 \times 2} = \frac{-5 \pm 1}{4}.$$

Hence the solutions are $x = -1$ and $x = \frac{-3}{2}$.

Direct substitution can be used to check that these are the correct solutions.

Setting $x = -1$ in the original equation:

$$2x^2 + 5x + 3 = 2(-1)^2 + 5(-1) + 3 = 2 - 5 + 3 = 0.$$

Setting $x = \frac{-3}{2}$ in the original equation:

$$2x^2 + 5x + 3 = 2\left(\frac{-3}{2}\right)^2 + 5\left(\frac{-3}{2}\right) + 3 = \frac{9}{2} - \frac{15}{2} + 3 = 0\cdot$$

Hence, both solutions are correct. (It is *always* worth checking your solution to a quadratic equation in this way.)

Example 2.2

Solve the equation $3x^2 - 10x + 1 = 0$ for x.

Solution

Setting $a = 3$, $b = -10$ and $c = 1$ in the quadratic equation formula, gives

$$x = \frac{10 \pm \sqrt{10^2 - 4 \times 3 \times 1}}{2 \times 3} = \frac{10 \pm \sqrt{88}}{2 \times 3}$$
$$= \frac{5 \pm \sqrt{22}}{3} = \frac{5 \pm 4.690}{3}.$$

Hence the solutions are $x = 3.23$ and $x = 0.103$.

These solutions can be checked as follows.

For $x = 3.23$: $3x^2 - 10x + 1 = 3 \times (3.23)^2 - 10 \times 3.23 + 1 = -0.001$.

For $x = 0.103$: $3x^2 - 10x + 1 = 3 \times (0.103)^2 - 10 \times 0.103 + 1 = 0.002$.

So both solutions are correct to within the accuracy with which we have carried out our calculations.

Method 2: Solution by factorization

This method essentially involves guessing the solutions, $x = \alpha$ and $x = \beta$ say, and then showing that they are correct. This may sound like a tall order, and it does require a certain amount of practice. However it is to some extent simplified by the fact that if α and β are the solutions of Equation 2.40 then that equation can be **factorized**, that is rewritten in the form

$$a(x - \alpha)(x - \beta) = 0. \qquad (2.42)$$

That α and β are solutions of this equation is clear, since setting x equal to either α or β will certainly force Equation 2.42 to be true (provided $a \neq 0$). The fact that Equation 2.42 is also equivalent to the general quadratic equation is seen by; (i) multiplying out the parentheses in Equation 2.42 to obtain

$$ax^2 + a(-\alpha - \beta)x + a\alpha\beta = 0$$

and (ii) by comparing this with Equation 2.40. Thus if Equations 2.40 and 2.42 are to be alternative ways of writing the same equation then α and β must be related to a, b and c by the equations

$$\alpha + \beta = \frac{-b}{a} \qquad (2.43)$$

and $$\alpha\beta = \frac{c}{a}. \qquad (2.44)$$

The real point of establishing these conditions for the equivalence of Equations 2.40 and 2.42 is that Equations 2.43 and 2.44 may help you to guess the solutions α and β in the first place, or at least to check that the solutions you have guessed are correct.

Example 2.3

Solve the equation $x^2 + 2x = 0$ by factorization.

Solution

This equation can be easily factorized, by writing it in the form $x(x + 2) = 0$. This shows that the two solutions are $x = 0$ and $x = -2$, (as claimed earlier).

Example 2.4

Solve the equation $2x^2 + 5x + 3 = 0$ by factorization.

Solution

This equation can be factorized, by noting that the sum of its solutions must be $-5/2$, and the product of its solutions must be $3/2$. The solutions themselves must therefore be $x = -1.5$ and $x = -1$, which allows the original quadratic equation to be written in the form $2(x + 1.5)(x + 1) = 0$.

Question 2.20 Show that the expression $(x - 1)(x + 2)$ can be written in the form $ax^2 + bx + c$, and find a, b and c. Hence solve $x^2 + x - 2 = 0$ for x. ■

Solving quadratic equations by factorization can be quicker than using the quadratic equation formula, but it requires a lot of practice and a certain amount of trial and error. If you are not already familiar with factorization, then it is probably better to use the formula method. In any case, it is certainly worth committing the formula (Equation 2.41) to memory; it is a frequently used result.

Question 2.21 Solve the equation $x^2 - x - 2 = 0$ for x.

Question 2.22 Solve the equation $y^2 - 6y + 1 = 0$ for y.

Question 2.23 Solve the equation $s^2 - 4s + 4 = 0$ for s.

Question 2.24 Solve the equation $x^2 - 1 = 0$. ■

4.7 Launching bodies from a height

Not all projectiles start and finish at the same height above the Earth's surface. In times gone by, a cannon-ball might have been fired from a castle wall or a cliff, and in modern warfare a bomb might be dropped from an aeroplane cruising at great height (Figure 2.27). Both the cannon-ball and the bomb are projectiles and both follow trajectories that are, to a first approximation, part of a parabola, but in both cases the launch point is well above the target point. Figure 2.28 illustrates a typical case of this sort in which a projectile is launched at an initial speed u and at an angle θ to the horizontal, from a height h above ground level. This is the kind of projectile motion that will be considered in this subsection.

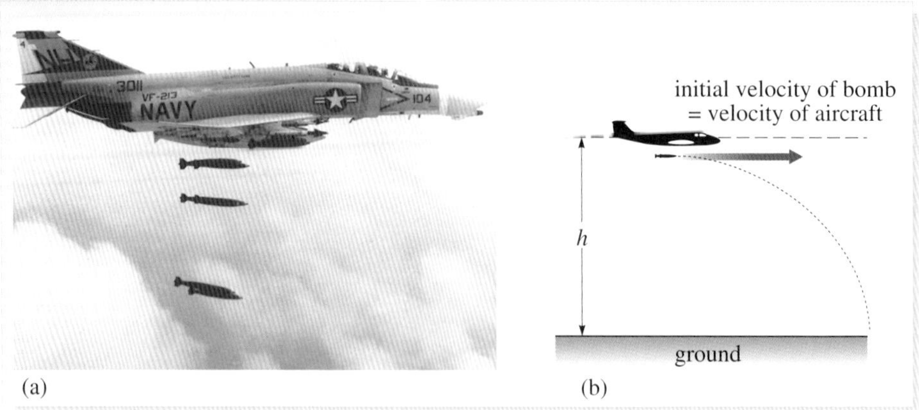

Figure 2.27 (a) Bombs falling from a warplane; this photograph was taken from another aircraft at the same altitude and travelling with the same velocity. (b) Viewed from the Earth each bomb follows the path of a projectile.

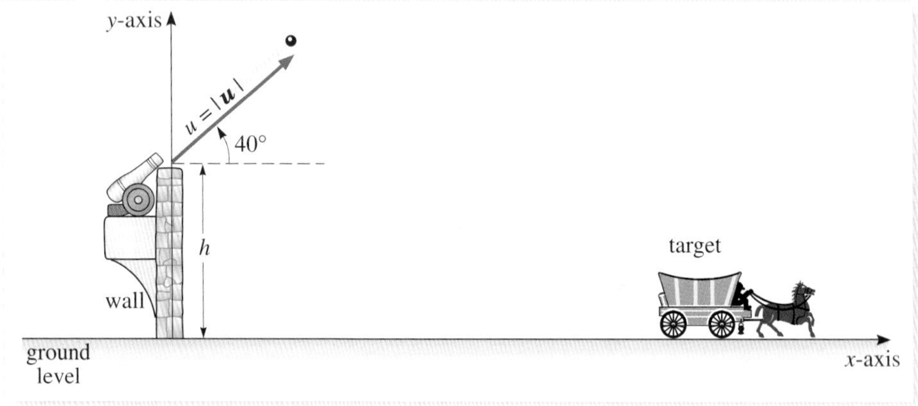

Figure 2.28 A projectile launched from a wall of height h with speed u, at an angle θ to the horizontal. The origin of coordinates is at the base of the wall, but the projectile's displacement is still measured from its initial position at the launch point.

In order to analyse the motion of a projectile the first step is to select a suitable coordinate system. In the case of Figure 2.28 we have chosen to use a conventional pair of orthogonal x- and y-axes, with their common origin at the base of the wall, but we will continue to measure displacements from the launch point. As before, we call the initial velocity components u_x and u_y, with $u_x = u\cos\theta$ and $u_y = u\sin\theta$, and we regard the horizontal and vertical components of motion to be independent apart from the fact that they have the same duration. The horizontal component of velocity will remain constant throughout the flight, and so the x-component of the displacement (from the launch point) will change at a steady rate.

So $$v_x = u_x \qquad \text{(Eqn 2.26)}$$

and $$s_x = u_x t. \qquad \text{(Eqn 2.27)}$$

The y-component of velocity will decrease at a steady rate due to the constant downward acceleration of gravity, which has magnitude g. So,

$$v_y = u_y - gt, \qquad \text{(Eqn 2.30)}$$

and, as a result, the y-component of the displacement from the launch point will be given by

$$s_y = u_y t - \tfrac{1}{2}gt^2. \qquad \text{(Eqn 2.31)}$$

All of this should be familiar. The only real difference from what we did before is that the target point is now at a distance h below the launch point, so the final vertical component of the projectile's displacement must be $s_y = -h$. (Note that h

itself is a distance, and therefore a positive quantity.) Substituting this value into Equation 2.31 shows that the time t at which the projectile arrives at the target point must satisfy the equation

$$-h = u_y t - \tfrac{1}{2} g t^2. \qquad (2.45)$$

This is a quadratic equation in the unknown t. If you are already experienced in dealing with quadratic equations this will be obvious, but if not you may find it helpful to rearrange Equation 2.45 thus

$$\tfrac{1}{2} g t^2 - u_y t - h = 0. \qquad (2.46)$$

Comparing this with Equation 2.40 you should be able to see that Equation 2.46 is a quadratic equation in which $a = g/2$, $b = -u_y$, and $c = -h$. Remembering that in this case the unknown has been called t, it follows from the quadratic equation formula that the solutions are

$$t = \frac{u_y + \sqrt{u_y^2 + 2gh}}{g} \quad \text{and} \quad t = \frac{u_y - \sqrt{u_y^2 + 2gh}}{g}.$$

The second of these solutions is physically unacceptable, since it is a negative quantity ($2gh$ is a positive quantity, so when the square root is evaluated it will turn out to be greater than u_y, implying that $u_y - \sqrt{u_y^2 + 2gh}$ will be negative). Hence, the projectile reaches the target at the time

$$t = \frac{u_y + \sqrt{u_y^2 + 2gh}}{g}. \qquad (2.47)$$

However the projectile was launched at $t = 0$ (this is when $s_y = 0$), so, in this case, the arrival time is also equal to the time of flight T, hence

$$T = \frac{u_y + \sqrt{u_y^2 + 2gh}}{g}. \qquad (2.48)$$

According to Equation 2.27, the horizontal component of the projectile's displacement at the end of its flight will be $s_x = u_x T$, i.e.

$$s_x = \frac{u_x u_y + u_x \sqrt{u_y^2 + 2gh}}{g}. \qquad (2.49)$$

It follows from this that the range of the projectile will be the positive quantity

$$R = \left| \frac{u_x u_y + u_x \sqrt{u_y^2 + 2gh}}{g} \right|. \qquad (2.50)$$

As in Section 4.5, the trajectory of the projectile can be found by eliminating t from Equations 2.27 and 2.31. The result is the same as before, a parabola described by the equation

$$s_y = u_y\left(\frac{s_x}{u_x}\right) - \frac{1}{2}g\left(\frac{s_x}{u_x}\right)^2. \qquad \text{(Eqn 2.38)}$$

At this stage you should already be asking yourself if Equations 2.48 and 2.50 make sense. For instance, if you set $h = 0$ do they reproduce the results we found for T and R in Sections 4.3 and 4.4? (They do!) Does a positive non-zero value of h alter the values of T and R in the way that you would expect; in particular, does it increase them or decrease them? (It increases them, as you should expect.) Always try to read equations in this way.

Question 2.25 The description of projectile motion given above is based on the behaviour of the projectile's displacement from its launch point, $\boldsymbol{s} = (s_x, s_y)$. However, it is almost as easy to describe the motion directly in terms of the projectile's position $\boldsymbol{r} = (x, y)$. Carry out this alternative analysis for yourself by:

(a) writing down the initial and final positions of the particle using the coordinate system of Figure 2.28;

(b) noting that Equations 2.26 and 2.30 still apply, and writing down the equations involving x and y that are equivalent to Equations 2.27 and 2.31;

(c) using the four equations discussed in part (b) to determine the time of flight, the range and the trajectory of the projectile. (Note that this should lead you to the equation for a parabola in terms of x and y rather than s_x and s_y.) ■

Example 2.5

A cannon-ball is fired upwards at an angle of 45° to the horizontal from the top of a cliff that is 100 m above sea-level. If the initial speed of the ball is $40.0\,\mathrm{m\,s^{-1}}$, what is its range?

Solution

In this case

$$u_x = u\cos\theta = (40.0\,\mathrm{m\,s^{-1}})\cos 45° = 28.3\,\mathrm{m\,s^{-1}}$$

$$u_y = u\sin\theta = (40.0\,\mathrm{m\,s^{-1}})\sin 45° = 28.3\,\mathrm{m\,s^{-1}}.$$

Substituting these results into Equation 2.50 gives

$$R = \left|\frac{(28.3\,\mathrm{m\,s^{-1}})^2 + (28.3\,\mathrm{m\,s^{-1}})\sqrt{(28.3\,\mathrm{m\,s^{-1}})^2 + 2(9.81\,\mathrm{m\,s^{-2}})(100\,\mathrm{m})}}{9.81\,\mathrm{m\,s^{-2}}}\right|$$

$$= \left|\frac{(28.3\,\mathrm{m\,s^{-1}})^2 + (28.3\,\mathrm{m\,s^{-1}})\sqrt{2762\,\mathrm{m^2\,s^{-2}}}}{9.81\,\mathrm{m\,s^{-2}}}\right|$$

$$= 233\,\mathrm{m}.$$

Example 2.6

A cannon-ball is fired upwards at an angle of 60° to the horizontal, from the top of a vertical wall, which is 30.0 m above the ground. The target is 100 m horizontally from the bottom of the wall. At what speed should the cannon-ball be fired?

Solution

What we need is an equation that relates the unknown speed u to known quantities such as the range (100 m), the angle of launch (60°) and the launch height (30 m). The range equation itself (Equation 2.50) is not useful here, so we use the description of the trajectory given by Equation 2.38. The fact that enables us to evaluate the initial speed u is that the final displacement of the projectile is $\boldsymbol{s} = (s_x, s_y) = (100\,\text{m}, -30\,\text{m})$. Substituting these values, along with $\boldsymbol{u} = (u_x, u_y) = (u\cos 60°, u\sin 60°)$ and $g = 9.81\,\text{m}\,\text{s}^{-2}$ into Equation 2.38 gives

$$-30.0\,\text{m} = \frac{u\sin 60°(100\,\text{m})}{u\cos 60°} - \frac{(9.81\,\text{m}\,\text{s}^{-2})(100\,\text{m})^2}{2u^2\cos^2 60°}$$

i.e. $$-30.0\,\text{m} = 173\,\text{m} - \frac{1.96\times 10^5\,\text{m}^3\,\text{s}^{-2}}{u^2}.$$

giving $$203\,\text{m} = \frac{1.96\times 10^5\,\text{m}^3\,\text{s}^{-2}}{u^2},$$

so that $$u^2 = \frac{1.96\times 10^5\,\text{m}^3\,\text{s}^{-2}}{203\,\text{m}}.$$

Since u is positive by definition, we choose the positive square root of this expression, which gives $u = 31.1\,\text{m}\,\text{s}^{-1}$.

Question 2.26 Suppose the situation in Example 2.6 is changed so that the cannon-ball is fired upwards at an angle of 30° to the horizontal, the height of the wall is 50 m and the target is 200 m horizontally from the base of the wall. Calculate the speed at which the cannon-ball should be fired in order to hit the target.

Question 2.27 A rifle bullet is fired horizontally at $300\,\text{m}\,\text{s}^{-1}$ from the top of a cliff which is 100 m above the sea. Find the horizontal distance travelled by the bullet before it hits the sea.

Question 2.28 A rifle bullet identical to the one mentioned in Question 2.27 is dropped from the top of the cliff at the same time that the rifle is fired horizontally. If air resistance is ignored, which of the bullets will strike the water first? ■

5 Motion in space

5.1 From two dimensions to three

The term **space** is used to mean the collection of all possible positions that a particle might have. As mentioned in Chapter 1, it is a fact of common experience that space is three-dimensional; there are three *independent* directions in which a particle can move in space, and any point in space can be specified by just three position coordinates.

So far in this chapter, our discussion has concentrated on motion in a plane. Such motion is restricted to two dimensions, which makes it easy to portray on a flat sheet of paper using just two perpendicular axes, usually labelled x and y. However, the time has now come to face up to the reality of three dimensions and to add a third axis to the coordinate system we use to describe motion. This third axis is usually called the z-axis.

Figure 2.29 shows one way of arranging three mutually perpendicular axes to form a three-dimensional coordinate system. Such a system is sometimes referred to as a three-dimensional **Cartesian coordinate system**, after the French mathematician and philosopher René Descartes (1596–1650) who made important contributions to the development of geometry by using coordinates. As usual, the coordinates are measured from the common point of intersection, which is called the *origin*, and are given positive values in the directions indicated by the arrowheads. The coordinates are measured in units of length, usually metres.

The particular arrangement shown in Figure 2.29 is consistent with our earlier use of the y-axis as the vertical axis, and it is the one we shall adopt for the rest of this chapter. However, you should be aware that there are many other possibilities. A more common arrangement, that you will meet later, is obtained by rotating the system shown in Figure 2.29 so that the z-axis is vertical while the x- and y-axes point in perpendicular directions in the horizontal plane. You should always feel free to rotate the coordinate system in this way in order to simplify a problem.

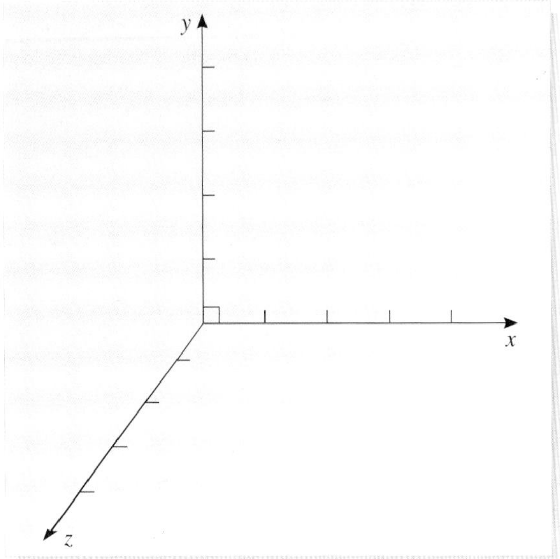

Figure 2.29 A three-dimensional coordinate system consists of three calibrated, mutually perpendicular axes, meeting at a unique origin. This particular arrangement of axes constitutes a right-handed coordinate system. The z-axis points towards you.

Another arrangement that is much less common, but which you should be aware of so that you can *avoid* using it, is illustrated in Figure 2.30. The systems shown in Figures 2.29 and 2.30 are related in the same way that your right hand is related to your left hand. There is no way of rotating your right hand so that it exactly matches your left (a left glove cannot be rotated to fit a right hand) but the mirror reflection of your right hand does match your left hand. In a similar way, the coordinate system shown in Figure 2.29, which is said to be a **right-handed coordinate system**, can never be rotated into alignment with the left-handed system shown in Figure 2.30, though its mirror image could be so aligned. It is a scientific convention to always use right-handed systems of coordinates. The way to recognize a right-handed system is to compare it with the first three fingers of your right hand as shown in Figure 2.31.

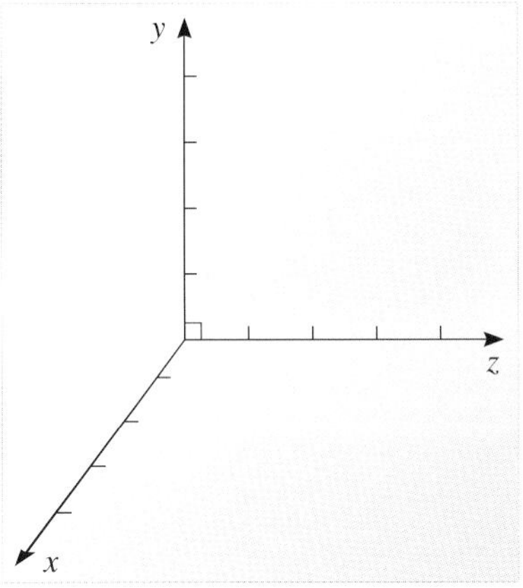

Figure 2.30 A left-handed coordinate system. This should be contrasted with the right-handed system shown in Figure 2.29. The x-axis points towards you.

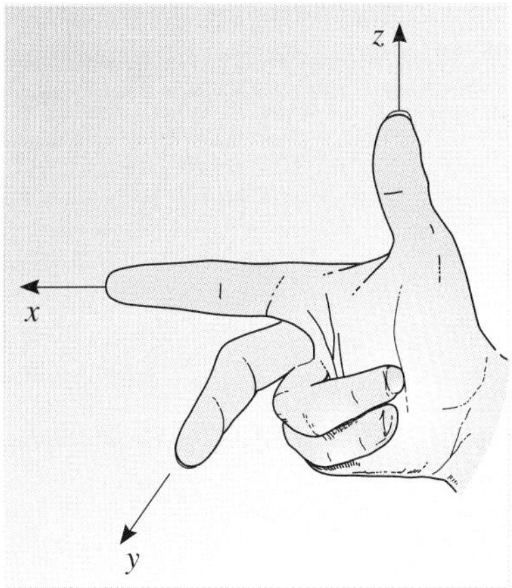

Figure 2.31 A test for recognizing right-handed coordinate systems. If you align the thumb and index finger of your right hand with the z- and x-axes respectively, then the y-axis will align with your second finger if the coordinate system is right-handed. You can use this test to confirm that the coordinate system in Figure 2.29 is right-handed, while that in Figure 2.30 is not.

5.2 Vectors in space

Having established a coordinate system, the position of any point in three-dimensional space is specified by its x-, y- and z-coordinates. Figure 2.32 shows how this is done, by constructing perpendiculars to each of the three axes. It is not easy to show the perpendiculars accurately on a perspective drawing; it would be better to work with a three-dimensional wire model. However, this practical difficulty is not important. What is important is that you understand how the x-, y- and z-coordinates of a point are determined in principle, so that you can interpret the equations in which they appear.

Drawing and interpreting diagrams that illustrate three-dimensional motions can also be difficult. Fortunately it is often possible to avoid relying on such diagrams by using vectors instead. We have already examined two-dimensional vectors in some detail, in Sections 2 and 3. In this subsection we generalize the ideas developed in those earlier sections to three dimensions and develop some additional ideas that are especially useful in three dimensions.

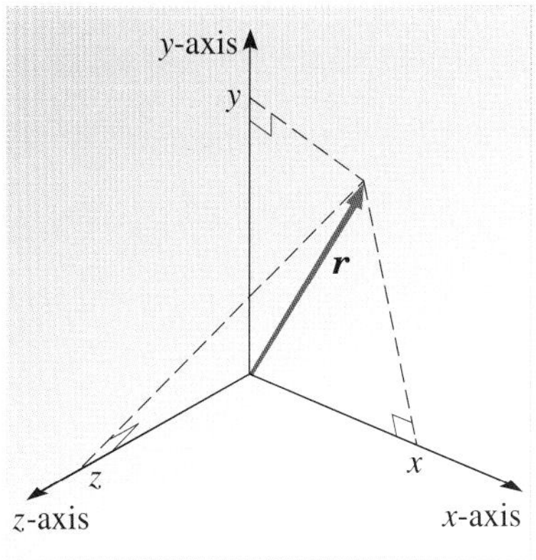

Figure 2.32 The coordinates x, y and z of a point are determined by constructing perpendiculars to the axes.

As in the two-dimensional case, we can represent the position vector of a point diagrammatically by means of an arrow from the origin to the point (Figure 2.32). We continue to denote the position vector of the point by the bold symbol $\boldsymbol{r}$, but $\boldsymbol{r}$ now has three components, usually called its **Cartesian components**, which are equal to the coordinates of the point. Consequently we can write:

$$\boldsymbol{r} = (x, y, z). \qquad (2.51)$$

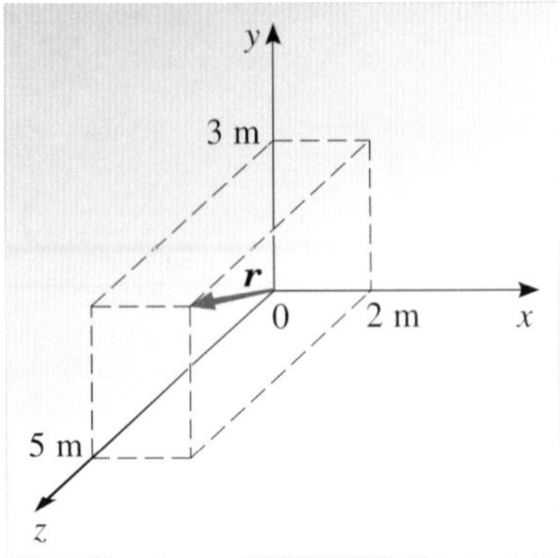

Figure 2.33 The position vector of the point with $x = 2$ m, $y = 3$ m and $z = 5$ m.

An example is shown in Figure 2.33. In this case the position vector is $\boldsymbol{r} = (2\,\text{m}, 3\,\text{m}, 5\,\text{m})$, and it is represented by an arrow that joins the origin to the point with coordinates $x = 2$ m, $y = 3$ m and $z = 5$ m. Notice that the order within the parentheses *is* important; (2 m, 3 m, 5 m) does *not* have the same meaning as (3 m, 5 m, 2 m); these are *ordered triples*.

The magnitude r of the position vector $\boldsymbol{r} = (x, y, z)$ is the distance from the origin to the point (x, y, z). This magnitude is sometimes written as $|\boldsymbol{r}|$ and is related to the coordinates by

$$r = |\boldsymbol{r}| = \sqrt{x^2 + y^2 + z^2}. \tag{2.52}$$

This is essentially the three-dimensional generalization of Pythagoras' theorem, which was introduced in Section 2.

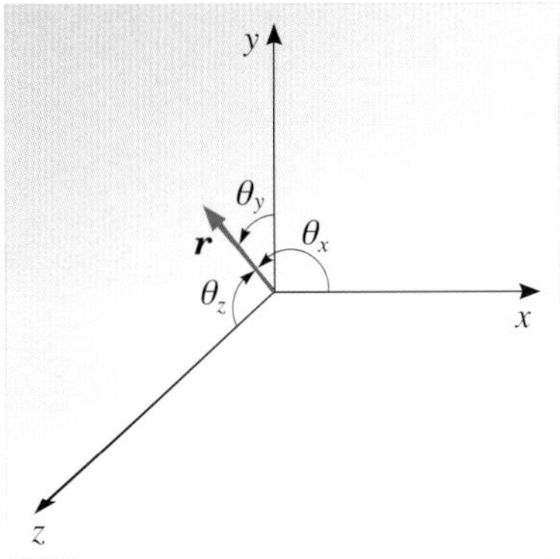

Figure 2.34 The x-, y- and z-coordinates, the distance r and the angles θ_x, θ_y and θ_z are all related.

Specifying directions in three dimensions can be quite difficult. Vectors provide a rather natural way of providing directional information, but if you wish you may also describe the direction of a vector by specifying the angles that it makes with the various coordinate axes. Figure 2.34 shows one way of doing this, by specifying the angles θ_x, θ_y and θ_z from the coordinate axes to the direction of $\boldsymbol{r}$. The components of the position vector $\boldsymbol{r}$ are related to the magnitude and direction of $\boldsymbol{r}$ as follows

$$x = r\cos\theta_x, \quad y = r\cos\theta_y, \quad z = r\cos\theta_z. \tag{2.53}$$

Although Figure 2.34 shows three angles, θ_x, θ_y and θ_z, it is important to realize that they are not all independent. The fact that they are related follows from Equations 2.52 and 2.53, which show that

$$r^2 = x^2 + y^2 + z^2 = r^2(\cos^2\theta_x + \cos^2\theta_y + \cos^2\theta_z) \tag{2.54}$$

hence $$\cos^2\theta_x + \cos^2\theta_y + \cos^2\theta_z = 1. \tag{2.55}$$

This relation implies that only two of the angles θ_x, θ_y, θ_z are independent. It provides a way of checking that three given angles are consistent, or of working out one angle when given the other two. For example, the angles $\theta_x = \theta_y = 60°$ and $\theta_z = 45°$ (used in Question 2.30) *are* consistent since, in that case

$$\cos^2\theta_x + \cos^2\theta_y + \cos^2\theta_z = 2\cos^2 60° + \cos^2 45°\cdot$$

Now $\cos 60° = 1/2$ and $\cos 45° = 1/\sqrt{2}$

so $$\cos^2\theta_x + \cos^2\theta_y + \cos^2\theta_z = 2\left(\frac{1}{2}\right)^2 + \left(\frac{1}{\sqrt{2}}\right)^2 = 1\cdot$$

Question 2.29 What is the magnitude of the position vector $\boldsymbol{r} = (2\,\text{m}, 3\,\text{m}, 5\,\text{m})$?

Question 2.30 A position vector has magnitude $r = 10$ m and the angles between that vector and the three axes of a Cartesian coordinate system are $\theta_x = \theta_y = 60°$ and $\theta_z = 45°$. What are the components of the position vector?

Question 2.31 Suppose the angles θ_x, θ_y, θ_z are all equal and that $r = 10$ m. (a) Find the angles, assuming they lie between 0° and 90°. (b) Find the components of the position vector $\boldsymbol{r}$. ■

Having defined position vectors in three dimensions it is a simple matter to define three-dimensional displacements. Displacements describe differences in position, so if $\boldsymbol{r}_1 = (x_1, y_1, z_1)$ and $\boldsymbol{r}_2 = (x_2, y_2, z_2)$ are the position vectors of the points P_1 and P_2 (see Figure 2.35), then the displacement $\boldsymbol{s} = (s_x, s_y, s_z)$ from P_1 to P_2 is

$$\boldsymbol{s} = \boldsymbol{r}_2 - \boldsymbol{r}_1 \tag{2.56}$$

or, in terms of components,

$$(s_x, s_y, s_z) = (x_2 - x_1, y_2 - y_1, z_2 - z_1). \tag{2.57}$$

The magnitude of this vector

$$s = \sqrt{s_x^2 + s_y^2 + s_z^2}$$

is the distance from P_1 to P_2.

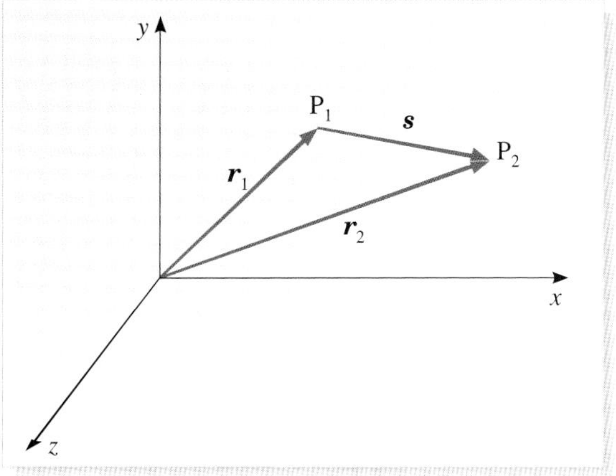

Figure 2.35 The displacement $\boldsymbol{s} = \boldsymbol{r}_2 - \boldsymbol{r}_1$.

Note that an equation relating three-dimensional vectors, such as Equation 2.56, is equivalent to three independent scalar equations $s_x = x_2 - x_1$, $s_y = y_2 - y_1$ and $s_z = z_2 - z_1$, corresponding to the three components of the original vector equation. This allows vector notation to be very compact.

If a particle is moving in three dimensions, then its position vector will change with time, as will its components, the position coordinates of the particle. Regarding the coordinates of the particle as functions of time, just as we did in the two-dimensional case in Section 3.1, we can define the three-dimensional instantaneous velocity $\boldsymbol{v} = (v_x, v_y, v_z)$ as the rate of change of the particle's position:

$$\boldsymbol{v} = \frac{\mathrm{d}\boldsymbol{r}}{\mathrm{d}t} = \left(\frac{\mathrm{d}x}{\mathrm{d}t}, \frac{\mathrm{d}y}{\mathrm{d}t}, \frac{\mathrm{d}z}{\mathrm{d}t}\right). \tag{2.58}$$

The magnitude of this vector,

$$v = \sqrt{v_x^2 + v_y^2 + v_z^2} \tag{2.59}$$

is equal to the (instantaneous) speed of the particle.

Acceleration follows a similar pattern. The instantaneous acceleration $\boldsymbol{a} = (a_x, a_y, a_z)$ is defined as the rate of change of the velocity, and is given by:

$$\boldsymbol{a} = \frac{d\boldsymbol{v}}{dt} = \left(\frac{dv_x}{dt}, \frac{dv_y}{dt}, \frac{dv_z}{dt}\right). \tag{2.60}$$

The magnitude of the acceleration is given by

$$a = \sqrt{a_x^2 + a_y^2 + a_z^2}\,. \tag{2.61}$$

Question 2.32 The position vector $\boldsymbol{r}$ of a particle is given by $\boldsymbol{r} = (at^2, bt, c)$, where t is the time, and a, b and c are constants. Giving your answers in terms of t, a, b and c, find expressions for (a) the velocity of the particle, (b) the speed of the particle and (c) the acceleration of the particle. ■

Three-dimensional vectors of a similar kind can be added together, and vectors may be multiplied by scalars. To add two vectors $\boldsymbol{u} = (u_x, u_y, u_z)$ and $\boldsymbol{v} = (v_x, v_y, v_z)$, for example, you just add their corresponding components

This rule for adding two vectors, Equation 2.62, is equivalent to the triangle rule.

$$\boldsymbol{u} + \boldsymbol{v} = (u_x + v_x, u_y + v_y, u_z + v_z) \tag{2.62}$$

and to multiply a vector $\boldsymbol{v} = (v_x, v_y, v_z)$ by a scalar λ, you just multiply each of its components by λ

$$\lambda\boldsymbol{v} = (\lambda v_x, \lambda v_y, \lambda v_z). \tag{2.63}$$

Of course, we cannot do something like add a velocity vector to an acceleration vector. Any terms that are to be added together must be expressible in the same units, as in $\boldsymbol{u} + \boldsymbol{a}t$.

5.3 Some examples of motion in space

The simplest form of motion in space is probably (three-dimensional) **uniform motion**, in which a particle travels with constant velocity (i.e. constant speed in a fixed direction). If the initial velocity of such a particle is $\boldsymbol{u} = (u_x, u_y, u_z)$, then the velocity $\boldsymbol{v}$ at any later time will be equal to the initial velocity, so

$$\boldsymbol{v} = \boldsymbol{u} \tag{2.64}$$

and, at time t, the particle's displacement from its initial position will be

$$\boldsymbol{s} = \boldsymbol{u}t. \tag{2.65}$$

If the particle's initial position (at $t = 0$) is denoted by the position vector $\boldsymbol{r}_0$, then the position at time t will be

$$\boldsymbol{r} = \boldsymbol{r}_0 + \boldsymbol{u}t. \tag{2.66}$$

As you can see, this is just a vector generalization of the description of uniform motion in one dimension that we examined in Chapter 1. In fact this is really exactly the same motion that we discussed in Chapter 1; the particle is still moving along a line, and the need to use vectors to describe it is simply a consequence of our desire to use a general Cartesian coordinate system rather than a system that has been carefully oriented to simplify the description.

A slightly more complicated kind of motion in three dimensions is **uniformly accelerated motion**. This too is described by vector generalizations of the corresponding one-dimensional equations. In this case, if the (constant) acceleration is $\boldsymbol{a} = (a_x, a_y, a_z)$, and the initial velocity of the particle is $\boldsymbol{u} = (u_x, u_y, u_z)$, then the velocity of the particle at time t will be

$$\boldsymbol{v} = \boldsymbol{u} + \boldsymbol{a}t \tag{2.67}$$

and the displacement of the particle from its initial position will be

$$\boldsymbol{s} = \boldsymbol{u}t + \tfrac{1}{2}\boldsymbol{a}t^2. \tag{2.68}$$

Once again, it should be noted that these two vector equations together actually provide six scalar equations.

Question 2.33 A particle undergoes uniformly accelerated motion with initial velocity $\boldsymbol{u} = (1, -10, 3)\,\text{m s}^{-1}$, and constant acceleration $\boldsymbol{a} = (2, 7, -4)\,\text{m s}^{-2}$.

(a) Find the velocity $\boldsymbol{v}$ of the particle after 2 s, and express that velocity in terms of its components. (b) What is the corresponding speed, v? (c) Find the displacement $\boldsymbol{s}$ from the initial position after 2 s and express that displacement in terms of its components. (d) How far is the particle from the origin after the first two seconds? ■

When discussing (two-dimensional) projectile motion great emphasis was given to the fact that the horizontal and vertical motions were independent *apart* from the fact that they had to have a common duration. This principle can be extended to three dimensions. We might for instance, consider a particle that moves with uniform acceleration in the y-direction and, simultaneously, with uniform motion in the xy-plane. Alternatively, we might consider a particle that moves with different uniform accelerations in the x- and y-directions, and with constant velocity in the z-direction. We might even consider a particle that has completely different uniform accelerations in all three directions or even non-uniform accelerations in all three directions. The ability to regard any three-dimensional motion as the result of combining three independent one-dimensional motions of equal duration is really nothing more than the recognition that space *is* three-dimensional and that the three-dimensional acceleration vector has three independent components.

Example 2.7

Consider a three-dimensional coordinate system set up with the xz-plane on a flat horizontal golf course, as shown in Figure 2.36. The z-axis points out of the paper at 90° to the xy-plane. The acceleration due to gravity acts in the negative y-direction. At time $t = 0$ s, a golfer hits a ball placed at the origin so that it sets off in the xy-plane at an angle of 45° to the x-axis. The initial speed of the ball is $35.0\,\text{m s}^{-1}$.

(a) Use the uniform acceleration equations to find (i) the time of flight of the ball; (ii) the coordinates of the point at which the ball should hit the course.

(b) Now assume that at time $t = 1.00$ s, a gust of wind gives the ball a constant component of velocity in the z-direction, equal to $2.00\,\text{m s}^{-1}$. Assuming the x- and y-components of the motion remain the same, find the coordinates of the point at which the ball hits the course.

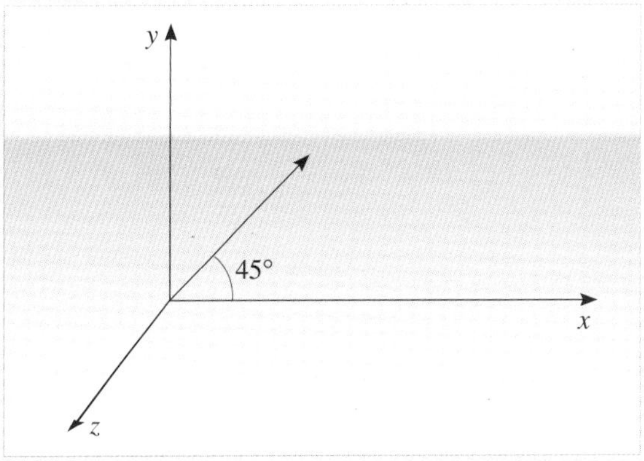

Figure 2.36 The coordinate system for Example 2.7.

Solution

(a) The fundamental equation for this problem is

$$\boldsymbol{s} = \boldsymbol{u}t + \tfrac{1}{2}\boldsymbol{a}t^2. \qquad \text{(Eqn 2.68)}$$

(i) To calculate the time of flight, we can use

$$s_y = u_y t + \tfrac{1}{2}a_y t^2 \qquad (2.69)$$

with $s_y = 0$ and $a_y = -g$, to obtain

$$0 = u_y t - \tfrac{1}{2}gt^2.$$

We are not interested in the $t = 0$ solution which represents the instant of projection. The other solution, representing the time when the ball lands on the course, is

$$t = \frac{2u_y}{g} = \frac{2(35\,\text{m}\,\text{s}^{-1})\sin 45°}{(9.81\,\text{m}\,\text{s}^{-2})}$$

$$= \frac{2 \times 35}{9.81 \times \sqrt{2}}\,\text{s} = 5.05\,\text{s}.$$

(ii) To find the x-component of displacement we use this value of t, together with $a_x = 0$, in Equation 2.68 to obtain

$$s_x = u_x t + \tfrac{1}{2}a_x t^2 = (35\,\text{m}\,\text{s}^{-1})\cos 45° \times (5.05\,\text{s})$$

$$= \frac{(35\,\text{m}\,\text{s}^{-1})(5.05\,\text{s})}{\sqrt{2}}$$

$$= 125\,\text{m}.$$

Since the ball is initially at the origin of coordinates we expect it to land at the position $\boldsymbol{r} = (125, 0, 0)\,\text{m}$.

Notice that if we use Equation 2.36 for the range, and assume that the motion remains in the xy-plane, then we obtain

$$R = \left|\frac{2u^2\cos\theta\sin\theta}{g}\right| = \frac{2 \times (35)^2}{9.81 \times (\sqrt{2})^2}\,\text{m} = 125\,\text{m}.$$

However, this approach requires remembering the (less fundamental) range equation, and is not what we were asked to do in the question! Moreover, we need the time of flight for the next part of the question.

(b) A similar calculation gives the z-component of the displacement, except that $u_z = 2\,\text{m}\,\text{s}^{-1}$ and $a_z = 0$. However, in this case, the time t must be measured from when the gust occurs, one second after the ball is struck. With this proviso

$$s_z = u_z t + \tfrac{1}{2}a_z t^2.$$

So, at the end of the flight, 4.05 s after the gust,

$$s_z = (2\,\text{m}\,\text{s}^{-1}) \times (4.05\,\text{s})$$

$$= 8.10\,\text{m}.$$

Again, since the ball started from the origin, the position where it lands will be $\boldsymbol{r} = (125, 0, 8.10)\,\text{m}$.

Question 2.34 Suppose the golfer described in Example 2.7 plays a second ball from the origin. The situation now is that at time $t = 0$ s, this ball is struck so that it sets off in the xy-plane at an angle of 60° to the x-axis. The initial speed of the ball is $40.0\,\mathrm{m\,s^{-1}}$.

(a) Use the uniform acceleration equations to find: (i) the time of flight of the ball; (ii) the coordinates of the point at which the ball should hit the course.

(b) Now assume that at time $t = 2.00$ s, a strong wind commences and continues to give the ball a constant component of acceleration in the z-direction, equal to $2.00\,\mathrm{m\,s^{-2}}$ throughout the remaining flight of the ball. Assuming the x- and y-components of the motion remain the same, find the coordinates of the point at which the ball hits the course. ■

6 Closing items

6.1 Chapter summary

1 A vector can be defined in two or three dimensions. In either case a vector has a magnitude and a direction. A scalar is specified by a single number, together with an appropriate unit of measurement. Position, displacement, velocity and acceleration are all examples of vector quantities. Mass, length, distance, speed and temperature are examples of scalar quantities.

2 Vectors are printed using bold letters $\boldsymbol{r}$, $\boldsymbol{s}$, $\boldsymbol{v}$, ... etc. and magnitudes are printed using italic letters r, s, v, ... or by using the magnitude symbol as in $|\boldsymbol{r}|$. In handwritten work, vectors should be indicated by a curly underline $\underset{\sim}{\mathrm{r}}$, $\underset{\sim}{\mathrm{s}}$, $\underset{\sim}{\mathrm{v}}$, ... and magnitudes written normally, with no underlining.

3 A two-dimensional vector $\boldsymbol{f}$, lying in the xy-plane can be specified in terms of its components by writing $\boldsymbol{f} = (f_x, f_y)$. The magnitude of such a vector is $f = \sqrt{f_x^2 + f_y^2}$, and the components are given by $f_x = f\cos\theta$ and $f_y = f\sin\theta$, where θ is the angle (measured anticlockwise) from the positive x-axis to the direction of $\boldsymbol{f}$.

4 Two vectors of the same type can be added together to form a resultant by adding their corresponding components

$$\boldsymbol{f} + \boldsymbol{g} = (f_x + g_x, f_y + g_y). \qquad (2.15)$$

The triangle rule provides a geometric interpretation of vector addition. (See Figure 2.12.)

5 A vector can be multiplied by a scalar λ (which is not necessarily an integer) by multiplying each of its components by λ

$$\lambda\boldsymbol{f} = (\lambda f_x, \lambda f_y). \qquad (2.16)$$

6 In three dimensions, the position vector $\boldsymbol{r}$ of a point P with position coordinates (x, y, z) is

$$\boldsymbol{r} = (x, y, z). \qquad (2.51)$$

The magnitude of this vector, $r = \sqrt{x^2 + y^2 + z^2}$, is the distance from the origin to the point P.

7 The displacement $\boldsymbol{s} = (s_x, s_y, s_z)$ from point P_1 with position vector $\boldsymbol{r}_1 = (x_1, y_1, z_1)$ to point P_2 with position vector $\boldsymbol{r}_2 = (x_2, y_2, z_2)$ is given, by

$$\boldsymbol{s} = \Delta\boldsymbol{r} = \boldsymbol{r}_2 - \boldsymbol{r}_1 = (x_2 - x_1, y_2 - y_1, z_2 - z_1) \tag{2.56}$$

The magnitude of this vector, $s = \sqrt{s_x^2 + s_y^2 + s_z^2}$, is the distance from P_1 to P_2.

8 The velocity $\boldsymbol{v} = (v_x, v_y, v_z)$ of a particle is determined by the rate of change of the particle's position

$$\boldsymbol{v} = \frac{\mathrm{d}\boldsymbol{r}}{\mathrm{d}t} = \left(\frac{\mathrm{d}x}{\mathrm{d}t}, \frac{\mathrm{d}y}{\mathrm{d}t}, \frac{\mathrm{d}z}{\mathrm{d}t}\right). \tag{2.58}$$

The magnitude of this vector, $v = \sqrt{v_x^2 + v_y^2 + v_z^2}$, represents the speed v of the particle.

9 The acceleration $\boldsymbol{a} = (a_x, a_y, a_z)$ of a particle is determined by the rate of change of the particle's velocity

$$\boldsymbol{a} = \frac{\mathrm{d}\boldsymbol{v}}{\mathrm{d}t} = \left(\frac{\mathrm{d}v_x}{\mathrm{d}t}, \frac{\mathrm{d}v_y}{\mathrm{d}t}, \frac{\mathrm{d}v_z}{\mathrm{d}t}\right). \tag{2.60}$$

10 In two (or three) dimensions the motion of a particle may be regarded as the sum of two (or three) separate motions, mutually at right angles, that are independent apart from their common duration.

11 A projectile is a body that moves in unpowered flight through the Earth's atmosphere under the influence of terrestrial gravity. If air resistance is ignored and the projectile is treated as a particle, then its motion is characterized by uniform motion in the horizontal direction (usually the x-direction) and by uniformly accelerated motion in the vertical direction (usually the y-direction). The motion of such a projectile is described by the equations

$$v_x = u_x \tag{2.26}$$

$$s_x = u_x t \tag{2.27}$$

$$v_y = u_y - gt \tag{2.30}$$

$$s_y = u_y t - \tfrac{1}{2} g t^2 \tag{2.31}$$

where g is the magnitude of the acceleration due to gravity, $\boldsymbol{u} = (u_x, u_y)$ is the initial velocity of the projectile and $\boldsymbol{s} = (s_x, s_y)$ is the displacement of the projectile from its initial position.

12 For a projectile launched over level ground, at an angle θ to the horizontal, with initial speed u, the time of flight is $T = (2u \sin\theta)/g$, the range is

$R = \left|\dfrac{2u^2 \sin\theta\cos\theta}{g}\right|$, and the trajectory is a parabola described by the equation

$$s_y = u_y\left(\frac{s_x}{u_x}\right) - \frac{1}{2}g\left(\frac{s_x}{u_x}\right)^2. \tag{2.38}$$

The range of such a projectile is a maximum when $\theta = 45°$.

13 For a projectile launched from a height h, the time of flight is

$$T = \frac{u_y + \sqrt{u_y^2 + 2gh}}{g}\text{, and the range is } R = \left|\frac{u_x u_y + u_x\sqrt{u_y^2 + 2gh}}{g}\right|.$$

14 A quadratic equation is an equation of the general form

$$ax^2 + bx + c = 0. \qquad (2.40)$$

Such an equation has two solutions, which are given by the quadratic equation formula as

$$x = \frac{-b \pm \sqrt{b^2 - 4ac}}{2a}. \qquad (2.41)$$

The two solutions will be identical if $b^2 = 4ac$. If the two solutions are $x = \alpha$ and $x = \beta$, then

$$\alpha + \beta = \frac{-b}{a} \quad \text{and} \quad \alpha\beta = \frac{c}{a}$$

and the original equation may be factorized to give

$$a(x - \alpha)(x - \beta) = 0. \qquad (2.42)$$

15 In three dimensions, Cartesian coordinate systems may be right-handed or left-handed. Either may be used when describing three-dimensional motion, though it is conventional to use a right-handed system.

16 Uniformly accelerated motion in three-dimensional space may be described by the equations

$$\boldsymbol{a} = \text{constant vector}$$

$$\boldsymbol{v} = \boldsymbol{u} + \boldsymbol{a}t \qquad (2.67)$$

$$\boldsymbol{s} = \boldsymbol{u}t + \tfrac{1}{2}\boldsymbol{a}t^2. \qquad (2.68)$$

6.2 Achievements

Now that you have completed this chapter, you should be able to:

A1 Explain the meaning of all the newly defined (emboldened) terms introduced in the chapter.

A2 Express vectors in terms of a magnitude and direction, or in terms of components.

A3 Write down and use the algebraic rules for adding vectors and for multiplying a vector by a scalar; also interpret and implement those rules graphically.

A4 Distinguish between position and displacement vectors and use both in the description of motion.

A5 Express the velocity and acceleration vectors of a particle in terms of the rate of change of its position and velocity vectors; also relate those vectors to the coordinates of the particle.

A6 Write down the principles and basic equations describing projectile motion. Use them to solve a variety of problems and to derive expressions for the time of flight, range and trajectory of a projectile in sufficiently simple circumstances.

A7 Recognize quadratic equations, solve them using the quadratic equation formula; factorize a quadratic equation given its solutions and recognize expressions relating the constants that appear in a given quadratic equation to the sum and the product of its solutions.

A8 Distinguish right-handed from left-handed Cartesian coordinate systems, and use right-handed systems in the description of three-dimensional motion with uniform acceleration.

6.3 End-of-chapter questions

Question 2.35 If the components of a two-dimensional velocity vector $\boldsymbol{v}$ are $v_x = 3\,\mathrm{m\,s^{-1}}$ and $v_y = 4\,\mathrm{m\,s^{-1}}$, find (a) the magnitude of $\boldsymbol{v}$, (b) the angle between the x-axis and the direction of $\boldsymbol{v}$.

Question 2.36 (a) Define the terms *position vector*, *velocity* and *acceleration*, in the context of a moving particle. (b) The position vector $\boldsymbol{r}$ of a particle is given by

$$\boldsymbol{r} = (At, 2At, 3At) + (3Bt^2, 2Bt^2, Bt^2)$$

where t is the time, and A and B are constants. Find the following in terms of A, B, and t: (i) the magnitude of the position vector, (ii) the components of the velocity, (iii) the components of the acceleration.

Question 2.37 Solve the following equations for x:

(a) $x^2 - 7x = 0$, (b) $x^2 - 7 = 0$, (c) $4x^2 + 4x - 3 = 0$.

Question 2.38 A motor-cycle stunt rider attempts to leap across a line of cars parked end-on, bumper to bumper. The rider accelerates directly up a ramp that is 50 m long and 10 m high at the take-off point, as shown in Figure 2.37. The rider takes off with a speed of $25\,\mathrm{m\,s^{-1}}$, so that the trajectory of the rider and the line of cars all lie in the same plane.
Throughout the calculations required for this question, you should assume that the ground is flat and horizontal, that air resistance is negligible, and that the motor cyclist may be treated as a particle. In parts (b) and (c), assume that the motor cyclist succeeds in his attempt to clear the cars.

(a) What is the maximum height above the ground that is attained by the motor cyclist? (*Hint*: At the point of maximum height, the vertical component of velocity is momentarily zero.)

(b) How long does the motor cyclist spend in the air?

(c) Calculate the speed of the motor cyclist at the instant of landing.

(d) During a trial run, the motor cyclist clears a line of 9 cars, each 4.50 m long. Would it be an acceptable risk for an tenth identical car to be added to the line before the stunt is performed again? (In addition to performing a calculation, you should comment on the validity of the assumptions made in this question.)

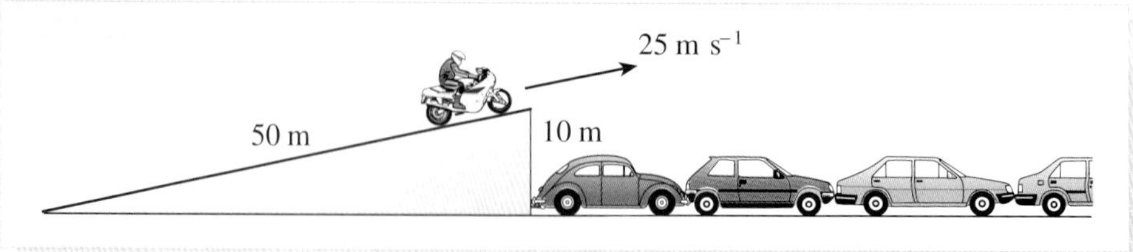

Figure 2.37 The situation described in Question 2.38.

Question 2.39 (a) Write down expressions that show how the components of the velocity and displacement of a projectile from the launch site vary with time. Explain the meaning of any symbols you use.

(b) A target moves at a constant speed u_T at a constant height h above a horizontal plane. At time $t = 0$, the target happens to fly directly over a projectile launching site which is located in the horizontal plane. At this instant, a projectile is launched at the target. Find an expression for $\tan\theta$ where θ is the angle at which the projectile must be launched so that, for the minimum possible launch speed, it just reaches the target. (You should give your answer in terms of u_T, h and the magnitude of the acceleration due to gravity.) ■

Chapter 3 Periodic motion

1 Earth satellites — an example of periodic motion

Artificial Earth satellites (commonly just called **satellites**) play a vital role in modern life (Figure 3.1). There are many kinds of satellite; communications satellites, weather satellites, Earth resources satellites, spy satellites and so on, but the one feature they all have in common — the thing that makes them *satellites* rather than *space probes* — is the fact that they move in closed *orbits* around the Earth. As shown in Figure 3.2, this means that each Earth satellite moves in a plane that passes through the centre of the Earth, repeatedly following a closed path of fixed shape and size. A satellite's orbital plane may be inclined at any angle to the plane of the Earth's equator, and its orbital path may have any one of a variety of shapes. However, from a practical point of view one particular orbit has an importance which outstrips that of all others. This very special orbit lies in the plane of the Equator (it is therefore said to be an *equatorial orbit*) and it takes the form of a circle with a radius of about 42 300 km. It is the orbit of most of the important communications satellites, including those used to relay satellite TV. This orbit is sometimes called the **Clarke orbit**, in honour of the British science-fiction author Sir Arthur C. Clarke who pointed out its significance in the 1940s, long before the first artificial satellites were launched.

Figure 3.1 Artificial Earth satellites. Top, the Hispasat telecommunications satellite. Middle, the Meteosat weather satellite. Bottom, the ERS-1 Earth resources satellite.

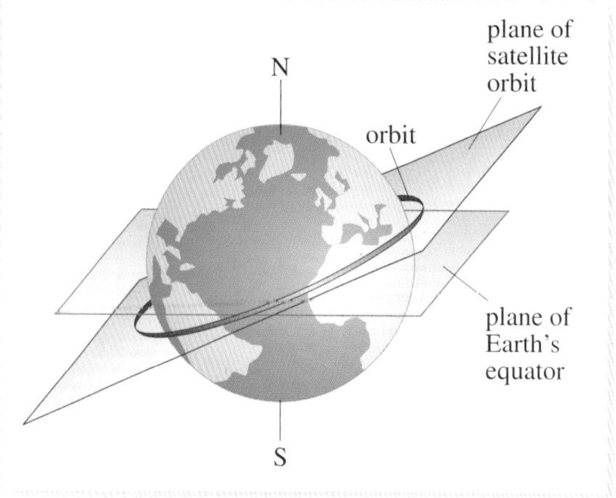

Figure 3.2 The orbit of an Earth satellite is a closed path in a plane that passes through the centre of the Earth. One of the basic characteristics of the orbit is the angle at which the orbital plane is inclined relative to the plane of the Earth's equator.

The special feature of the Clarke orbit that makes it so important is that any satellite moving in that particular equatorial orbit takes exactly 24 hours to circle the Earth, so when viewed from the Earth the satellite will appear to remain stationary above a fixed point on the Equator. Such a satellite is said to be **geostationary**. The great advantage of this, from the point of view of those responsible for satellite communications systems, is that the satellite appears to be stationary in the sky, so it can always stay in contact with a fixed geographical area (called the *footprint* of the satellite) and anyone wishing to receive signals from such a satellite only needs to aim their receiving dish at the appropriate *fixed* point in the sky (see Figure 3.3). Whenever you see a domestic satellite TV dish, it is almost certain to be directed towards some point on the Clarke orbit, which generally takes the form of a fixed arc in the sky, a bit like an invisible, but ever-present rainbow.

Figure 3.3 From any point on Earth, the Clarke orbit appears as a fixed arc in the sky. Satellite dishes are usually pointed at one of the geostationary satellites located on the arc. (Photo courtesy of BT.)

A satellite in orbit provides a good example of a body executing *periodic motion*. The term **periodic motion** refers to any kind of recurring motion that repeats itself after a characteristic interval of time called its **period**, usually denoted T. In the case of the Clarke orbit, the motion is circular and the period is $T = 24$ h. The mathematical feature that characterizes periodic motion is that whatever the position $\boldsymbol{r}(t)$ of a particle may be at time t, the particle will have the same position one period later, at time $t + T$. Thus, for periodic motion

$$\boldsymbol{r}(t + T) = \boldsymbol{r}(t) \quad \text{(for all values of } t\text{)} \tag{3.1}$$

and the period of the motion is the shortest time T that satisfies this relation. You should be able to think of several examples of motion that satisfy this condition, or at least come close to doing so for a while. The motions of planets and pendulums are good examples, as is the rotation of the hands of a clock or watch. Less 'pure' examples would be the motion of your head while nodding agreement, or the movement of your hand while shaking hands with a visitor. Life is full of periodic motions, without them we would have little sense of the regularity of the physical world and it is conceivable that science might never have arisen.

In this chapter we shall consider the description and significance of some of the commonest and most important kinds of periodic motion. We start with *uniform circular motion*, we then consider a common kind of oscillatory motion called *simple harmonic motion*, and finally we consider the *orbital motion* of planets and satellites of which circular orbits are an important special case. As in earlier chapters we shall pay particular attention to any general mathematical points that can be drawn from our physics-based discussions. In this case you will learn more about describing position coordinates and position vectors, but the main mathematical lesson will concern the so-called *periodic functions* that can be used to describe all kinds of periodic motion.

2 Circular motion

2.1 Some examples of circular motion

Circular motion is important in many areas of physics. This is because it is both easy to analyse and it provides a good approximation to many naturally occurring motions (Figure 3.4). The motion of the Moon about the Earth is an example of an approximately circular motion. Even more strictly circular motions are common in engineering; many of the moving parts of a motor vehicle execute a precise circular motion, and petrol engines often have the specific function of converting a linear motion (caused by burning petrol in a cylinder) into a circular rotary motion that can be used to make wheels turn. Also, as explained elsewhere in *The Physical World*, one of the earliest attempts to understand the internal structure of atoms involved the assumption that electrons moved in circles around the central nucleus of every atom.

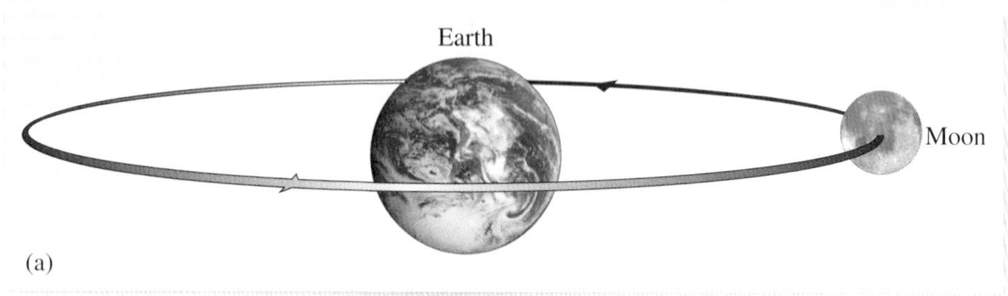

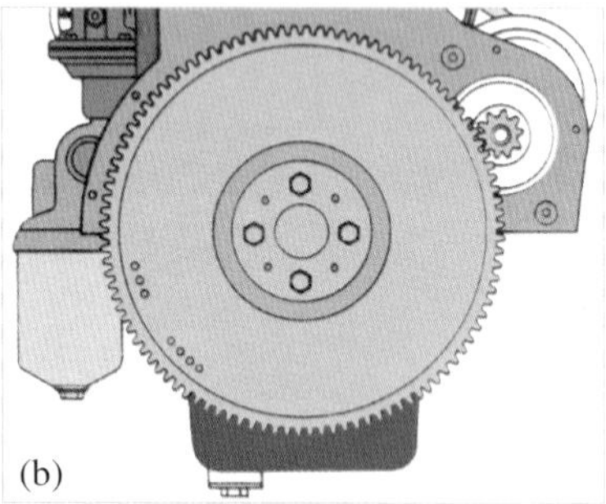

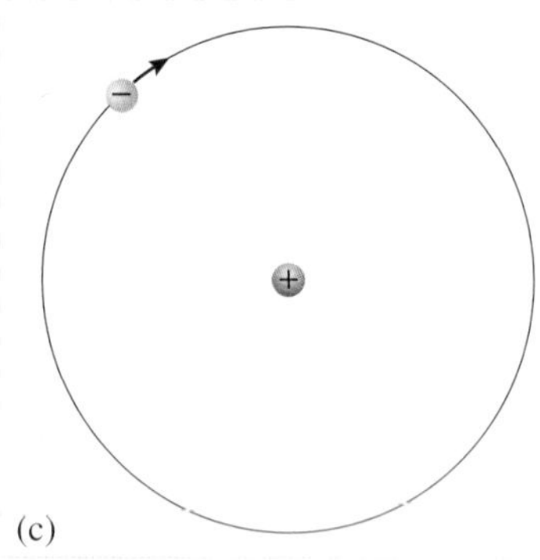

Figure 3.4 Some examples of circular motion: (a) the motion of the Moon, (b) a flywheel, (c) the circular motion of an electron in a crude model of an atom.

Circular motion, accurate or approximate, is common in both the natural and the technological world. Learning to describe such motion is mathematically interesting and scientifically essential — let's get on with the job.

2.2 Positions on a circle

As you will have gathered by now, the first step in analysing any kind of motion is to establish a suitable system of coordinates for the description of positions. In the case of motion in a circle there are two systems that make the description particularly simple. One is a two-dimensional *Cartesian system* in which the motion is in the *xy*-plane, and the origin is at the centre of the circle. The other is an entirely different system, called a *plane polar system*. We shall discuss the descriptions furnished by these two different systems in turn, starting with the plane polar system.

Plane polar coordinate description

In order to establish a system of **plane polar coordinates** it is first necessary to choose a point that can serve as the origin of the system and it is then necessary to choose a direction from the origin that can serve as a reference direction from which angles can be measured. Both these choices are arbitrary and should be made in such a way that they simplify the problem in hand as much as possible. Suitable choices for the description of motion in a circle are shown in Figure 3.5; the origin has been chosen to be at the centre of the circle and the reference direction is the rightward pointing axis. Once these choices have been made, any point in the plane with position vector $\boldsymbol{r}$, may be described by the ordered pair of plane polar coordinates $[r, \theta]$, where:

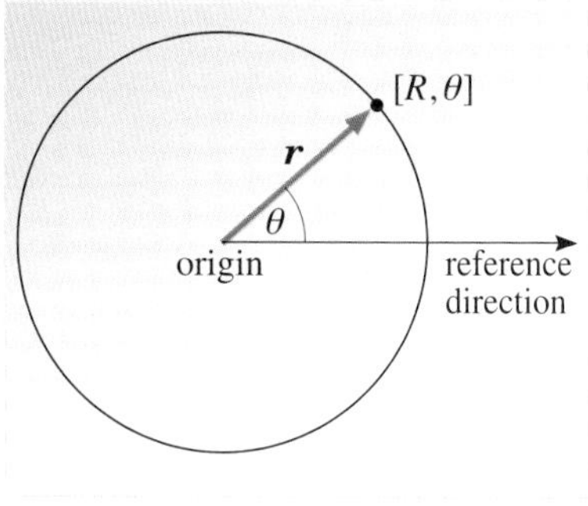

Figure 3.5 A system of plane polar coordinates for describing the position of a point on a circle of radius R.

r is the **radial coordinate** of the point, and indicates the distance from the origin to the point, i.e. the magnitude of the position vector $\boldsymbol{r}$;

θ is the **angular coordinate** of the point, and indicates the angle, measured in the anticlockwise direction, from the reference direction to the direction of the point, i.e. the direction of $\boldsymbol{r}$.

Using this system, any point on a circle of radius R, centred on the origin will have radial coordinate $r = R$, so its position vector $\boldsymbol{r}$ can be represented by $[R, \theta]$, where θ is the angular coordinate of the point. (Notice that the ordered pair of plane polar coordinates representing a point has been enclosed in square brackets; this is done to distinguish such a pair from an ordered pair of Cartesian coordinates, which would normally be enclosed in round brackets.)

An important point about plane polar coordinates that deserves comment concerns units. The radial coordinate r of a point must be measured in appropriate units of length, and these will usually be metres (m). Similarly, the angular coordinate θ, must be measured in some appropriate angular units. These might be degrees (°), but in many scientific applications it is more conventional and more convenient to use a different unit called the **radian** which may be represented by the symbol rad. The radian is quite a large angular unit; in fact, 1 rad = 57.296° (to three decimal places), so a radian is only a little less than 60°, but the exact relation that really defines a radian is the following

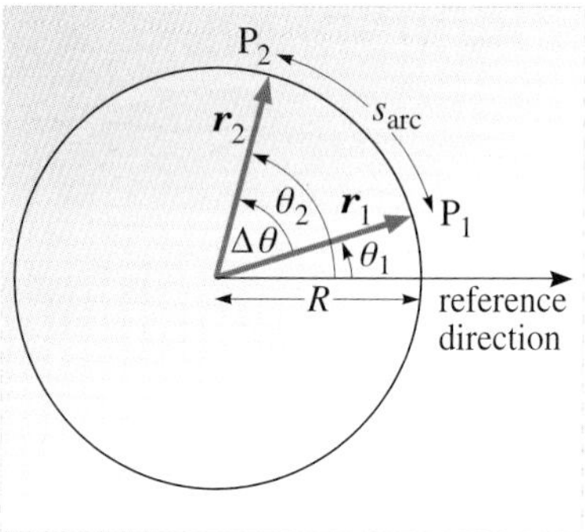

Figure 3.6 The arc length s_{arc} between two points on the circumference of a circle of radius R.

$$2\pi \text{ rad} = 360° \tag{3.2}$$

where π is the mathematical constant that relates the circumference C of a circle to its radius R (via $C = 2\pi R$), and which has the numerical value $\pi = 3.142$ (to three decimal places).

One advantage of using the radian as the unit of angle is that it makes it very easy to calculate the lengths of circular arcs. For instance, consider the points P_1 and P_2 on the circle of radius R shown in Figure 3.6. The points have respective position vectors $\boldsymbol{r}_1$ and $\boldsymbol{r}_2$ which may be represented in terms of plane polar coordinates by writing

$$\boldsymbol{r}_1 = [R, \theta_1] \quad \text{and} \quad \boldsymbol{r}_2 = [R, \theta_2].$$

It follows that the **angular displacement** from P_1 to P_2 is $\Delta\theta = \theta_2 - \theta_1$. If the angular coordinates have been measured in radians (rather than degrees or any other angular unit) then $\Delta\theta$ will also be expressed in radians and it will be found that the **arc length** (i.e. the length measured *along* the arc) separating P_2 from P_1 will be

$$s_{\text{arc}} = R\,|\Delta\theta| \quad (\Delta\theta \text{ in radians}). \tag{3.3}$$

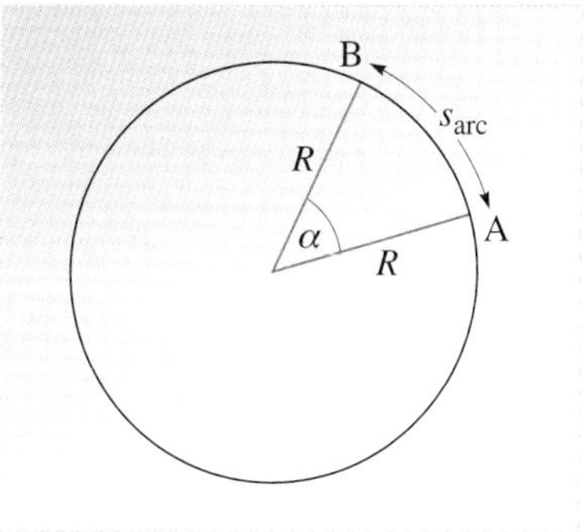

Figure 3.7 Two points A and B on the circumference of a circle of radius R, separated by arc length s_{arc}, subtend an angle $\alpha = s_{\text{arc}}/R$ at the centre of the circle.

For example, if $|\Delta\theta| = 2\pi$ then $s_{\text{arc}} = 2\pi R$, the circumference of the circle. Note that the arc length s_{arc} is always a positive quantity but $\Delta\theta$ may be positive or negative, so the right-hand side of Equation 3.3 has to involve the modulus $|\Delta\theta|$, which is always positive. An immediate consequence of Equation 3.3 is that if P_1 and P_2 are separated by 1 radian, then the arc length between them will be R; this is the case that has been illustrated in Figure 3.6.

A simple result that generalizes Equation 3.3 is that if A and B are two points on the circumference of a circle of radius R as in Figure 3.7, and if the arc length that separates them is s_{arc}, then the (positive) angle between A and B, as seen from the centre of the circle, will be

$$\alpha = s_{\text{arc}}/R \quad (\alpha \text{ in radians}). \tag{3.4}$$

α is the Greek letter alpha.

The arc length s_{arc} is said to **subtend** the angle α at the centre of the circle.

Note that Equations 3.3 and 3.4 only work if $\Delta\theta$ and α are measured in radians and even then, if s_{arc} is expressed in the same units as R it is necessary to disregard the angular unit in the product $R\,|\Delta\theta|$ and to introduce such a unit in the ratio s_{arc}/R. For instance, if $R = 5$ m, and $|\Delta\theta| = 2$ rad, then $R\,|\Delta\theta|$ is 10 m rad, but s_{arc} is just 10 m;

the rad is simply discarded. Treating units in this way is usually quite wrong, but if you regard Equation 3.4 as *defining* the angle α you can see that α is a purely numerical quantity, a dimensionless ratio that does not need a unit. In the view of many physicists this justifies neglecting radians whenever they become an embarrassment, and you will often see examples of this. It *is* possible to treat radians 'properly', in just the way that we treat metres or seconds, but in order to do so it is also necessary to treat some common equations in a rather unconventional way. For instance, Equation 3.4 would have to be written in the form $\alpha/\mathrm{rad} = s_{\mathrm{arc}}/R$. Rather than do this we will simply ignore or introduce a rad in a product of units whenever it is necessary or conventional to do so.

Question 3.1 Use the fact that 2π radians is equivalent to 360° to show that one radian is about 57.3°.

Question 3.2 (a) Express the following angles in terms of degrees: (i) $\pi/2$ rad, (ii) $\pi/3$ rad, (iii) $\pi/4$ rad, (iv) π rad.

(b) Sketch a circle of radius R, and on it show the points with the following plane polar coordinates: (i) $[R, \pi/2\ \mathrm{rad}]$, (ii) $[R, \pi/3\ \mathrm{rad}]$, (iii) $[R, \pi/4\ \mathrm{rad}]$, (iv) $[R, \pi\ \mathrm{rad}]$.

(c) If $R = 2$ m, what is the arc length separating the points with plane polar coordinates $[R, \pi/2\ \mathrm{rad}]$ and $[R, \pi/3\ \mathrm{rad}]$? ■

Cartesian coordinate description

A two-dimensional system of Cartesian coordinates, centred on a circle of radius R, is shown in Figure 3.8. Every point on the circumference of the circle must be at a distance R from the origin, so it follows from Equation 2.2 that any point on the circle will have Cartesian coordinates (x, y) such that

$$x^2 + y^2 = R^2. \qquad (3.5)$$

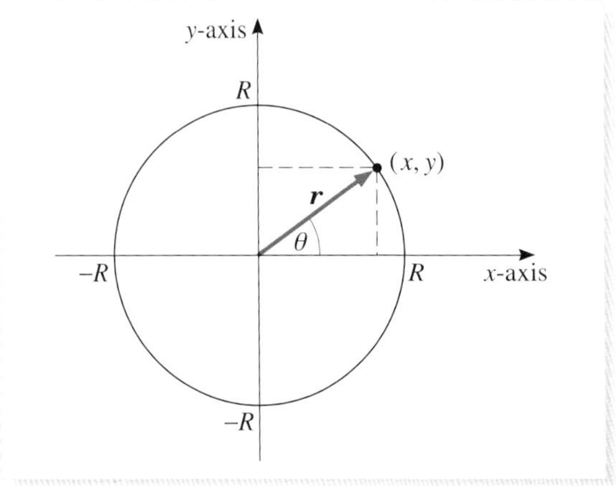

Figure 3.8 A system of Cartesian coordinates for describing the position of a point on a circle of radius R.

Equation 3.5 is said to be the **equation of a circle**. Every point on the circle satisfies this equation, and every point that satisfies this equation is on the circle. The equation is the algebraic representation of the circle. It is important to realize that the values of x and y that satisfy Equation 3.5 may be negative or positive. For instance, while it is certainly the case that $(x, y) = (R/\sqrt{2},\ R/\sqrt{2})$ satisfies Equation 3.5 and is therefore a point on the circle in Figure 3.8, it is also true that each of the points $(R/\sqrt{2}, -R/\sqrt{2})$, $(-R/\sqrt{2}, R/\sqrt{2})$ and $(-R/\sqrt{2}, -R/\sqrt{2})$ also satisfies Equation 3.5 and each is therefore also a point on the circle.

It is useful to know that Equation 3.5 describes the circle in Figure 3.8, but it would be even more useful to know general expressions for all the values of x and y that satisfy this equation. It is easy to write down such expressions in terms of the angle θ between the positive x-axis and the position vector $\boldsymbol{r}$ of a point, provided that θ is less than 90°. In such cases we can simply write

$$(x, y) = (R\cos\theta, R\sin\theta) \tag{3.6}$$

where $\cos\theta$ and $\sin\theta$ are the *trigonometric ratios* that were introduced in Chapter 2. But what happens when θ is greater than 90°, or when θ is less than 0°? Trigonometric ratios are not defined in such cases, but Equation 3.6 may still be used since it is possible to generalize the definitions of $\sin\theta$ and $\cos\theta$ so that they become meaningful for *all* values of θ yet still agree with the usual trigonometric ratios for angles in the range 0° to 90°. How this generalization is carried out is explained in Box 3.1 on *trigonometric functions*.

Box 3.1 Trigonometric functions

Figure 3.9 shows a point P with Cartesian coordinates (x, y) located on the circumference of a **unit circle** that is centred on the origin. A unit circle is a circle of radius 1. (This is a mathematical concept rather than a physical one, so the radius is not 1 m, or even 1 unit, it's just 1.) The coordinates x and y of point P will also be pure numbers in this case, so neither they, nor the axes along which they are measured have any associated units. The angular coordinate of P, measured anticlockwise from the positive x-axis, is θ.

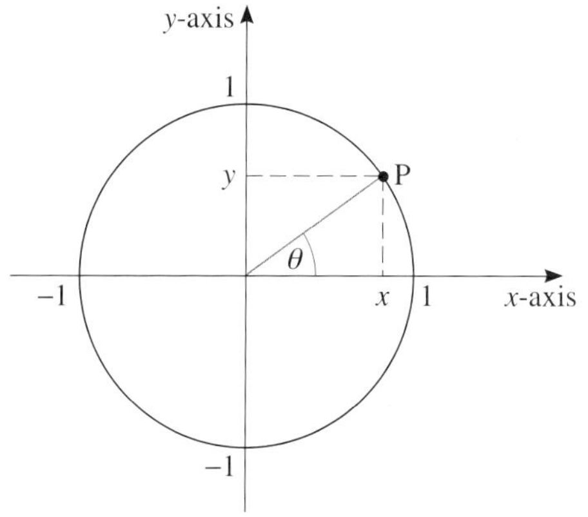

Figure 3.9 The coordinates x and y of a point P on a unit circle centred on the origin.

If P moves around the unit circle in an anticlockwise direction, starting from the point (1, 0) on the positive x-axis, then θ will increase from 0° and the values of x and y will change as it does so. As long as θ is less than 90° the coordinates of P will be $x = \cos\theta$ and $y = \sin\theta$, but even when θ exceeds 90° the coordinates x and y will remain well-defined and could be measured from a modified version of Figure 3.9 (i.e. from a version in which the appropriate values of x and y are used in place of the particular values shown in Figure 3.9). Even when θ exceeds 360° and P begins a second circuit of the unit circle, the coordinates x and y remain well defined; they simply repeat the same cycle of values they went through during the first circuit. Similarly, if P moves around the unit circle in the clockwise direction, starting from (1, 0), θ will decrease from 0° becoming more negative with time. However, it is still the case that well-defined values of x and y may be associated with every negative value of θ, no matter how large or small it may be. Thus, by using appropriately modified versions of Figure 3.9, it is possible to associate particular values of x and y with *any* value of θ, positive or negative. In view of this, we can *define* the **trigonometric functions**, sine(θ) and cosine(θ), usually abbreviated to sin(θ) and cos(θ), in terms of Figure 3.9, by the equations

$$\sin(\theta) = y \quad \text{for } any\ \theta \tag{3.7}$$

$$\cos(\theta) = x \quad \text{for } any\ \theta. \tag{3.8}$$

The graphs of these two functions are shown in Figures 3.10 and 3.11, respectively.

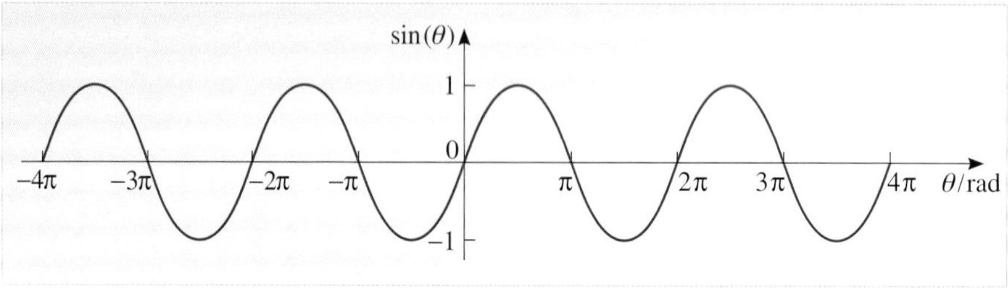

Figure 3.10 The graph of sin(θ) plotted against θ.

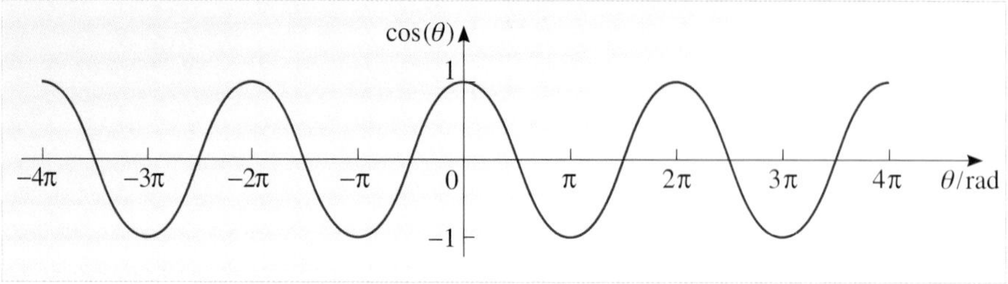

Figure 3.11 The graph of cos(θ) plotted against θ.

Note that in plotting the graphs we have chosen to measure θ in radians. This ensures that whatever value the function may have at one particular value of θ, the same function values will recur whenever θ is increased by 2π rad. We can describe this behaviour in mathematical terms by writing

$$\sin(\theta + 2\pi) = \sin(\theta) \quad \text{for } any\ \theta \text{ in radians} \tag{3.9}$$

$$\cos(\theta + 2\pi) = \cos(\theta) \quad \text{for } any\ \theta \text{ in radians.} \tag{3.10}$$

The repetitive nature of sin(θ) and cos(θ) means that they are **periodic functions**, while Equations 3.9 and 3.10 show that their period (of repetition) is 2π. Of course, an increase in θ by 2π rad corresponds to a complete circuit by P and hence to a full cycle of function values; this explains why sin($\theta + 2\pi$) *must* equal sin(θ), and why cos($\theta + 2\pi$) *must* equal cos(θ).

The trigonometric functions sin(θ) and cos(θ) have many important mathematical properties. One relates to the symmetry of Figures 3.10 and 3.11 as they spread out on either side of the origin. If you examine the graphs carefully you will see that cos (θ) is symmetric about the vertical axis but sin(θ) is not. In fact

$$\sin(-\theta) = -\sin(\theta) \quad \text{for } any\ \theta \tag{3.11}$$

but $$\cos(-\theta) = \cos(\theta) \quad \text{for } any\ \theta. \tag{3.12}$$

This behaviour is summarized by saying that sin(θ) is an **odd function** of θ (since the value of the function changes sign when θ is replaced by $-\theta$), while cos(θ) is an **even function** of θ (since its value is unaffected when θ is replaced by $-\theta$).

Another important property is a consequence of the fact that $x^2 + y^2 = 1$ for points on the unit circle. Since sin(θ) = y and cos(θ) = x, it follows that

$$\sin^2(\theta) + \cos^2(\theta) = 1 \quad \text{for } any\ \theta. \tag{3.13}$$

For certain special values of their **argument** θ, the trigonometric functions have particularly simple values. These are given in Table 3.1. For θ in the range 0 rad to $\pi/2$ rad the table repeats information that was given in Chapter 2, but here the results are extended to cover a complete period, from 0 rad to 2π rad. You can extend the results given in the table to even larger angles by using Equations 3.9 and 3.10. For angles that are less than 0 rad, you can achieve a similar extension by using Equations 3.11 and 3.12.

Table 3.1 Values of the sine and cosine functions for certain arguments. The values at 2π are not shown, but are identical to those at zero.

θ/rad	sin(θ)	cos(θ)	θ/rad	sin(θ)	cos(θ)
0	0	1	π	0	-1
$\pi/6$	$\frac{1}{2}$	$\frac{\sqrt{3}}{2}$	$7\pi/6$	$\frac{-1}{2}$	$\frac{-\sqrt{3}}{2}$
$\pi/4$	$\frac{1}{\sqrt{2}}$	$\frac{1}{\sqrt{2}}$	$5\pi/4$	$\frac{-1}{\sqrt{2}}$	$\frac{-1}{\sqrt{2}}$
$\pi/3$	$\frac{\sqrt{3}}{2}$	$\frac{1}{2}$	$4\pi/3$	$\frac{-\sqrt{3}}{2}$	$\frac{-1}{2}$
$\pi/2$	1	0	$3\pi/2$	-1	0
$2\pi/3$	$\frac{\sqrt{3}}{2}$	$\frac{-1}{2}$	$5\pi/3$	$\frac{-\sqrt{3}}{2}$	$\frac{1}{2}$
$3\pi/4$	$\frac{1}{\sqrt{2}}$	$\frac{-1}{\sqrt{2}}$	$7\pi/4$	$\frac{-1}{\sqrt{2}}$	$\frac{1}{\sqrt{2}}$
$5\pi/6$	$\frac{1}{2}$	$\frac{-\sqrt{3}}{2}$	$11\pi/6$	$\frac{-1}{2}$	$\frac{\sqrt{3}}{2}$

It sometimes happens that you know the value of sin(θ) or cos(θ), and you need to determine the corresponding value of θ. Problems of this kind can be solved by using the *inverse trigonometric functions* arcsine and arccosine (usually abbreviated to arcsin and arccos). For example, you can see from Table 3.1 that arcsin($1/\sqrt{2}$) = $\pi/4$ and arccos($\sqrt{3}/2$) = $\pi/6$. (On some calculators, arcsin is labelled $\sin^{-1}$ and arccos is labelled $\cos^{-1}$.) The tangent function defined by tan(θ) = sin(θ)/cos(θ), where cos(θ) $\neq$ 0, is sometimes useful. Its inverse function is denoted arctan or $\tan^{-1}$.

As was pointed out in Chapter 2, your calculator is programmed to evaluate trigonometric functions, so you should have no difficulty in working out their values for any θ, provided you know how to operate your calculator. Here are some questions to test this.

Question 3.3 (a) Use your calculator to evaluate cos 630° and sin 630°. Use your results to find the Cartesian coordinates (x, y) of the point on the circle with equation $x^2 + y^2 = 25\ \text{m}^2$ that is 630° (measured anticlockwise) from the positive x-axis.

(b) Carry out a similar evaluation for the same circle, but this time for a point that is −9.20 rad from the positive x-axis. (Note that the angle is negative in this case, implying that it is measured in the clockwise direction, and take care to ensure that you have switched your calculator to radian mode before inputting the value −9.20. If necessary consult your calculator's instruction manual for advice on how to switch between degree mode and radian mode.)

Question 3.4 Choose any reasonable value for θ and use it to check Equations 3.9, 3.10, 3.11, 3.12 and 3.13 using your calculator. Note that θ must be expressed in radians for Equations 3.9 and 3.10 to be valid. ■

2.3 Angular velocity and angular speed

ω is the Greek letter omega.

As a particle moves around a circle of radius R, centred on the origin, its radial coordinate r remains constant (and equal to R), but its angular coordinate θ changes with time. An important quantity that characterizes these changes is the **angular velocity** of the moving particle. This is a vector quantity, usually denoted by the symbol $\boldsymbol{\omega}$, that requires both a magnitude and a direction for its complete specification.

The magnitude of $\boldsymbol{\omega}$ is usually represented by ω and is a positive quantity. It indicates the magnitude of the rate of change of θ, and is defined by the equation

$$\omega = \left|\frac{\mathrm{d}\theta}{\mathrm{d}t}\right|. \tag{3.14}$$

By convention the direction of $\boldsymbol{\omega}$ is always at right angles to the plane of the motion. So, if the motion is in the xy-plane, the angular velocity will be in the z-direction. However, this requirement does not completely determine the direction, since $\boldsymbol{\omega}$ might point in the positive z-direction or the negative z-direction. Its direction in any particular case depends on the *sense* of the motion, i.e. whether it is clockwise or anticlockwise, and is determined by the **right-hand grip rule** as explained in Figure 3.12. For the anticlockwise motion shown in Figure 3.12, $\boldsymbol{\omega}$ points upwards, but for a clockwise motion it would point downwards. (It should be clearly understood that this is a convention, justified only by the necessity for all physicists to mean the same thing when describing angular velocity.)

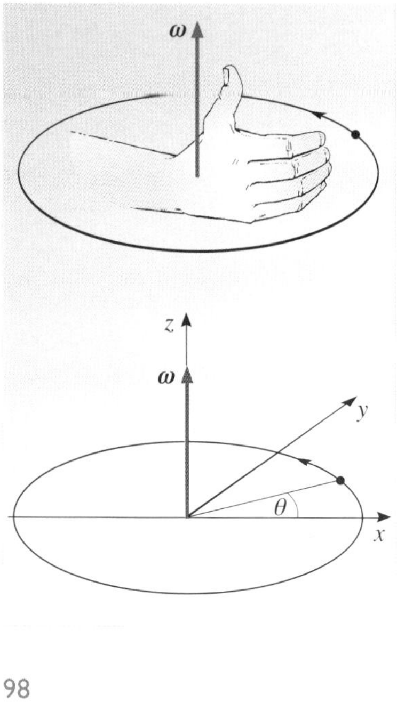

Figure 3.12 The right-hand grip rule that determines the direction of the angular velocity vector. Stretch the fingers of your right hand along the radius of the circle and then curl them in the sense of rotation of the particle. If necessary, turn your hand over to achieve the required alignment. The angular velocity vector is perpendicular to the plane of the orbit and in the direction of your outstretched thumb (rather than in the opposite direction).

The angular velocity vector $\boldsymbol{\omega}$ is useful because it gives a concise summary of the rate of rotation, the plane of rotation and the sense of rotation. Its magnitude ω is also an important quantity; it is called the **angular speed** of the motion and its SI unit is the radian per second (usually symbolized by rad s^{-1}). Note that the use of the term *angular speed* to describe the magnitude of an angular velocity is consistent with our use of the terms 'speed' and 'velocity' in Chapters 1 and 2.

In general, the variation of θ with time might be quite complicated. The circling particle might speed up or slow down (altering the magnitude of $\boldsymbol{\omega}$) or it might even reverse direction (altering the direction of $\boldsymbol{\omega}$). However, in what follows we shall avoid such complications by restricting our attention to the special case of **uniform circular motion**, in which the particle always moves in the same sense (always clockwise, or always anticlockwise) and does so in such a way that θ changes at a constant rate. This is the rotational equivalent of the uniform linear motion that we considered in Chapter 1.

Since we are interested in *uniform* circular motion, with θ changing at a constant rate, it follows that the derivative $\mathrm{d}\theta/\mathrm{d}t$ will be constant for any particular case we consider. If $\mathrm{d}\theta/\mathrm{d}t$ is positive, θ increases with time and the particle moves around the origin in the anticlockwise sense. If $\mathrm{d}\theta/\mathrm{d}t$ is negative, θ decreases with time and the particle moves in the clockwise sense. The equations describing uniform circular motion with radius R are therefore

$$\omega = \left|\frac{\mathrm{d}\theta}{\mathrm{d}t}\right| = \text{constant} \tag{3.15}$$

$$r = R \tag{3.16}$$

and, depending on the sense of the rotation

either $\quad \theta(t) = \omega t + \theta_0 \quad$ (for anticlockwise rotation) $\quad$ (3.17a)

or $\quad \theta(t) = -\omega t + \theta_0 \quad$ (for clockwise rotation) $\quad$ (3.17b)

where $\theta(t)$ represents the angular coordinate of the particle at any time t, and θ_0 represents the angular coordinate of the particle at $t = 0$, so $\theta_0 = \theta(0)$. (Equations 3.17a and 3.17b are the rotational analogues of Equation 1.6a: $x = v_x t + x_0$.)

It follows from Equations 3.6 and 3.17 that the Cartesian coordinates of a particle moving uniformly in a circle of radius R, centred on the origin, will be

$$(x, y) = (R\cos(\pm\omega t + \theta_0), R\sin(\pm\omega t + \theta_0)) \tag{3.18}$$

where the + signs describe anticlockwise motion and the − signs describe clockwise motion. We shall consider this description in more detail in Section 3.

Determining the angular speed ω of a uniform circular motion is straightforward. The rate of change of angular coordinate, $\mathrm{d}\theta/\mathrm{d}t$, is constant and may be determined by choosing any time interval Δt, measuring the corresponding change $\Delta\theta$, and then working out the ratio $\Delta\theta/\Delta t$. The angular speed will be given by the modulus of this ratio, so

$$\omega = \left|\frac{\mathrm{d}\theta}{\mathrm{d}t}\right| = \left|\frac{\Delta\theta}{\Delta t}\right|. \tag{3.19}$$

As usual, modulus symbols have been included here because $\Delta\theta/\Delta t$ might be negative, but ω must be positive. (Remember, $\mathrm{d}\theta/\mathrm{d}t$ represents the gradient of a graph of θ against t. If $\mathrm{d}\theta/\mathrm{d}t$ is constant, the graph will be a straight line and its gradient will be given by $\Delta\theta/\Delta t$.)

If T is the time period required for a particle to complete one full revolution of a uniform circular motion, then the angular coordinate θ of the particle will increase or decrease by 2π rad during that period, so $|\Delta\theta| = 2\pi$ rad when $|\Delta t| = T$. It follows from Equation 3.19 that for uniform circular motion with period T

$$\omega = \frac{2\pi\,\text{rad}}{T}. \tag{3.20}$$

For instance, every point on the Equator takes 24 h to circle the centre of the Earth, so the angular speed of each such point must be

$$\omega = \left|\frac{\Delta\theta}{\Delta t}\right| = \frac{2\pi\,\text{rad}}{24\,\text{h}} = \frac{2\pi\,\text{rad}}{24\times 60\times 60\,\text{s}} = 7.3\times 10^{-5}\,\text{rad}\,\text{s}^{-1}.$$

Question 3.5 The Moon orbits the Earth every lunar month (27 days and 7 hours). (a) Calculate its angular speed. (b) How long does the Moon take to move across the field of view of a telescope with an angular view of 1°? (Assume that allowance has already been made for the motion of the Earth.) ■

In addition to its angular velocity and angular speed, any particle moving in a circle also has an instantaneous velocity and an instantaneous speed in the usual (linear) sense that we discussed in Chapters 1 and 2. The rest of this subsection is devoted to the relationship between these angular and linear quantities.

Figure 3.13 shows a rotating disc. If the disc turns with constant angular speed ω every point on the disc will be engaged in uniform circular motion around the centre of the disc. Every point on the disc will have the same angular speed ω, but the instantaneous *linear* speed v of a point will depend on its distance from the centre of the disc. During one complete revolution, points near the rim of the disc have to travel a greater distance than points near the centre; they therefore have to travel at a higher (linear) speed.

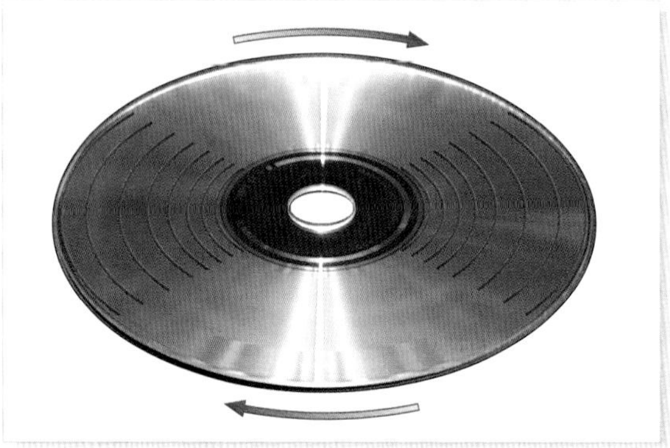

Figure 3.13 A rotating disc. The length of the red lines indicates increasing speed as distance from the centre increases.

If a point is at a distance r from the centre then the distance that it travels in one revolution is $2\pi r$, and the time that this takes is the period of revolution T. However, the linear speed is constant throughout this motion (even though the direction of motion is changing), so it is given by

$$v = \frac{2\pi r}{T}. \tag{3.21}$$

Now, Equation 3.20 tells us that $T = 2\pi/\omega$, and using this to eliminate T from Equation 3.21 we see that

$$v = r\omega. \tag{3.22}$$

This implies that for a fixed angular speed ω, the linear speed v simply increases in proportion to the distance r from the centre of the disc. Of course, the presence of the disc is not crucial to the correctness of Equation 3.22; the result applies equally well to a particle moving in a circle of radius r with angular speed ω, whether or not it is part of a rigid body.

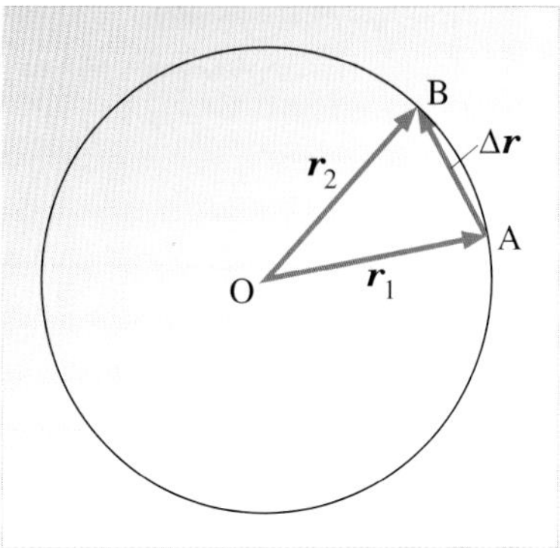

Figure 3.14 Two positions of a particle moving in a circle of radius r. At time t_1, the particle is at point A and has position vector $\boldsymbol{r}_1$, and at time t_2, it is at point B and has position vector $\boldsymbol{r}_2$.

It's worth noting that although a particle moving uniformly in a circle has a constant linear speed v, its instantaneous linear velocity $\boldsymbol{v}$ certainly isn't constant. The magnitude of $\boldsymbol{v}$ is constant, but its direction is changing all the time. To find the direction of $\boldsymbol{v}$ at any instant, consider Figure 3.14 which shows the position vector $\boldsymbol{r}_1$ of a circling particle at time t_1 and the position vector $\boldsymbol{r}_2$ of the same particle at a slightly later time, $t_2 = t_1 + \Delta t$. The displacement vector $\Delta \boldsymbol{r}$, representing the *change* in position, is also shown on the diagram. The particle's average velocity over the interval from t_1 to t_2 is given by

$$\langle \boldsymbol{v} \rangle = \frac{\Delta \boldsymbol{r}}{\Delta t}. \tag{3.23}$$

The direction of $\langle \boldsymbol{v} \rangle$ will always be the same as that of $\Delta \boldsymbol{r}$, but, as you can see from Figure 3.15, if Δt is reduced, so that point B gets closer to A, then the direction of the average linear velocity will get closer to being tangential to the circle. Since the instantaneous velocity $\boldsymbol{v}$ at time t_1 is equal to the average velocity in the limit as Δt approaches zero, we can say that:

> For a particle in circular motion, the instantaneous linear velocity is always tangential to the circle. If the circle is centred on the origin, the particle's velocity vector $\boldsymbol{v}$ will always be perpendicular to its position vector $\boldsymbol{r}$.

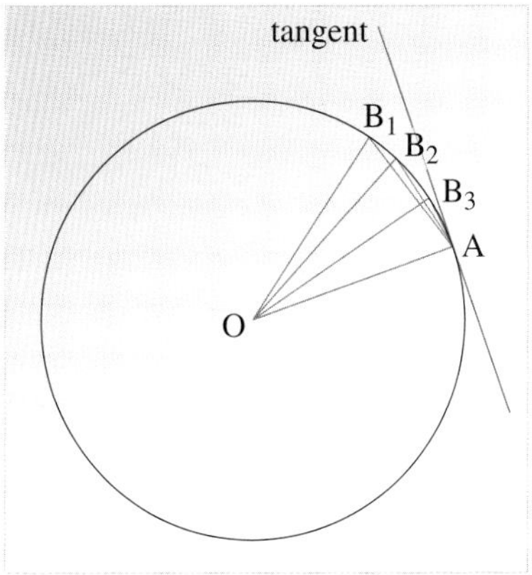

Figure 3.15 As the time interval becomes smaller and smaller, the direction of the average velocity gets closer and closer to being tangential to the circle.

This is illustrated in Figure 3.16, which shows the instantaneous velocity of a circling particle at two points along its pathway. To get some feeling for the truth of this result, imagine you are whirling a small object around your head, on the end of a piece of string. What would happen if the string suddenly snapped? The answer is that the object would fly off at a tangent to the circle — in the direction of its instantaneous velocity at the moment the string snapped.

Question 3.6 Use your answer to Question 3.5 to estimate the Moon's linear speed as it orbits the Earth. The distance between the Earth and the Moon may be taken to be 3.84×10^8 m.

Question 3.7 We are all carried around an approximate circle by the Earth's uniform rotation about its axis. The city of Entebbe is on the Equator (within a kilometre or two). Calculate how fast the citizens of Entebbe are moving as a consequence of the Earth's rotation. Take the radius of the Earth to be 6.38×10^6 m. ■

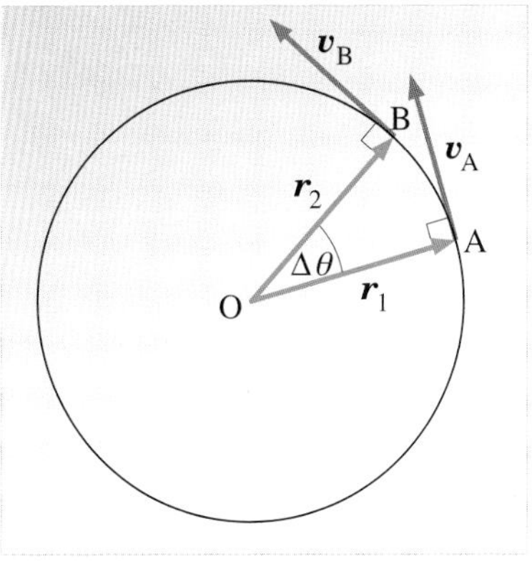

Figure 3.16 An object undergoing uniform circular motion. The velocity vectors are shown at positions A and B.

2.4 Centripetal acceleration

The velocity of a particle moving uniformly in a circle changes continuously. The magnitude is constant, but the direction is altering all the time. The fact that the velocity is changing means that the particle is accelerating, even though its speed is constant. We will now work out this acceleration, which is known as the **centripetal acceleration**.

Note the use of angle brackets to denote the average value of the quantity enclosed.

By definition, the average acceleration that occurs when the particle in Figure 3.16 moves from A to B during a time Δt is given by

$$\langle \boldsymbol{a}_{AB} \rangle = \frac{\Delta \boldsymbol{v}_{AB}}{\Delta t} \quad (3.24)$$

where $\Delta \boldsymbol{v}_{AB}$ is given by

$$\Delta \boldsymbol{v}_{AB} = \boldsymbol{v}_B - \boldsymbol{v}_A. \quad (3.25)$$

The centripetal acceleration will be equal to the limiting value of this average acceleration as Δt tends to zero.

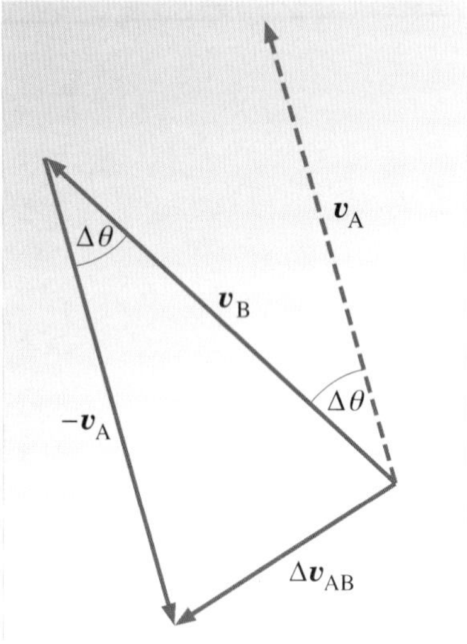

Figure 3.17 Using the triangle rule to evaluate $\Delta \boldsymbol{v}_{AB} = \boldsymbol{v}_B - \boldsymbol{v}_A = \boldsymbol{v}_B + (-\boldsymbol{v}_A)$.

The vector difference $\Delta \boldsymbol{v}_{AB}$ can be evaluated using the triangle rule for adding vectors (introduced in Chapter 2), as shown in Figure 3.17. The important point to notice is that the difference of two vectors, $\boldsymbol{v}_B - \boldsymbol{v}_A$, may be regarded as the *sum* of $\boldsymbol{v}_B$ and the vector $-\boldsymbol{v}_A$, where $-\boldsymbol{v}_A$ has the same magnitude as $\boldsymbol{v}_A$, but points in the opposite direction. Note also that the angle $\Delta\theta$ between $\boldsymbol{v}_A$ and $\boldsymbol{v}_B$ in Figure 3.17 is the same as the angle between OA and OB in Figure 3.16. This is because $\boldsymbol{v}_A$ is at a right angle to OA, and $\boldsymbol{v}_B$ is at a right angle to OB. If the position vector sweeps through $\Delta\theta$, from OA to OB, then the instantaneous velocity must sweep through the same angle. We can use Figure 3.17 to find the magnitude and direction of the centripetal acceleration.

The direction of the centripetal acceleration

As Δt tends to zero, $\Delta\theta$ also tends to zero and the angle between $\Delta \boldsymbol{v}_{AB}$ and $\boldsymbol{v}_A$ tends to a right angle. Since $\boldsymbol{v}_A$ is in the direction of the tangent to the circle at A, this means that $\Delta \boldsymbol{v}_{AB}$ must point towards the centre of the circle as $\Delta\theta$ tends to zero. It follows that the centripetal acceleration must also be directed towards the centre of the circle. Since there is nothing special about the point A, this is a general result.

In uniform circular motion, the centripetal acceleration always acts radially inwards, towards the centre of the circle.

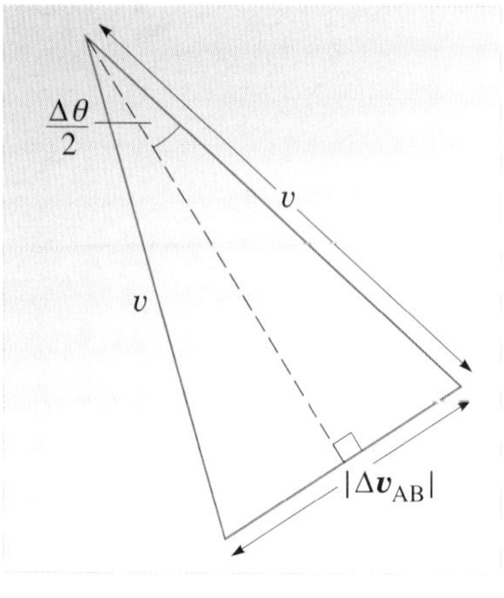

Figure 3.18 The isosceles triangle of Figure 3.17.

The magnitude of the centripetal acceleration

We first remark that, for small angles (measured in radians), the sine of the angle is approximately equal to the angle itself:

$$\sin(\theta) \approx \theta \quad (\theta \text{ small and in radians}). \quad (3.26)$$

You can check this with your calculator, but remember that θ must be in radians for the approximation to work. You will see that Equation 3.26 works well for θ less than 0.2 rad or about 12°.

The triangle shown in Figure 3.17 is an isosceles triangle since it has two sides of equal length. (Both $\boldsymbol{v}_A$ and $\boldsymbol{v}_B$ are of magnitude v.) It follows (see Figure 3.18) that

$$|\Delta \boldsymbol{v}_{AB}| = 2v\sin(\Delta\theta/2) \approx v\,\Delta\theta \quad (3.27)$$

However, from Equation 3.19 we have

$$\Delta\theta = \omega\Delta t \quad (\text{for positive } \Delta\theta) \quad (3.28)$$

$$\text{therefore} \quad |\Delta \boldsymbol{v}_{AB}| \approx v\omega\Delta t. \quad (3.29)$$

As Δt (and hence $\Delta \boldsymbol{v}_{AB}$) becomes smaller, Equation 3.29 becomes an increasingly accurate approximation, implying that the magnitude of the average acceleration,

$|\Delta \boldsymbol{v}_{AB}/\Delta t|$ approaches the value $v\omega$. In the limit as Δt tends to zero the relationship becomes exact and we get the following important expression for the magnitude of the centripetal acceleration

$$a = v\omega. \tag{3.30}$$

Since $v = r\omega$, we can also write the magnitude of the centripetal acceleration in the alternative forms

$$a = r\omega^2 \tag{3.31}$$

and
$$a = \frac{v^2}{r}. \tag{3.32}$$

Question 3.8 A wheel with a diameter of 80 cm is spinning so that a point on its perimeter has a speed of $30\,\text{km}\,\text{h}^{-1}$. What is the acceleration of a point on the perimeter of the wheel?

Question 3.9 The Moon's angular speed was worked out in Question 3.5, where it was found to be $2.66 \times 10^{-6}\,\text{rad}\,\text{s}^{-1}$. Given that the Moon is at a distance of about 3.84×10^8 m, calculate the magnitude of the centripetal acceleration of the Moon in its orbit. ■

2.5 Circular Earth orbits

In Question 3.9 you were asked to calculate the centripetal acceleration of the Moon in its orbit. But what causes this acceleration? The answer is the Earth's gravity. The Moon, like all other satellites, natural or artificial, is essentially falling under gravity, but its instantaneous linear velocity is such that its path is almost circular so that it always manages to miss the Earth. Figure 3.19 gives you some idea of how this comes about. It shows schematically how the parabolic path of a projectile is modified when we take account of the Earth's curvature, and the fact that the acceleration due to gravity is always directed towards the centre of the Earth.

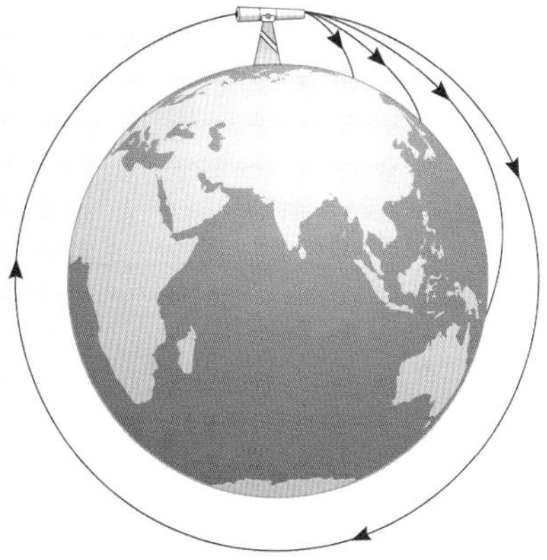

Figure 3.19 A projectile launched horizontally with a relatively low speed will fall back to the Earth, but one launched with a sufficiently high speed will fall around the Earth.

Near to the surface of the Earth, the acceleration due to the Earth's gravity has a magnitude of about $9.81\,\text{m}\,\text{s}^{-2}$, but in the neighbourhood of the Moon, as Question 3.9 shows, that magnitude is only about $2.73 \times 10^{-3}\,\text{m}\,\text{s}^{-2}$. The reason for this decline, as Sir Isaac Newton realized, is that gravity obeys an **inverse square law**.

This means that the magnitude of the acceleration due to gravity at a distance r from the centre of the Earth is given approximately by

$$a = \frac{\text{constant}}{r^2} = \frac{4.0 \times 10^{14}\ \text{m}^3\,\text{s}^{-2}}{r^2} \quad (\text{for } r \geq R_{\text{Earth}}) \tag{3.33}$$

where the symbol $\geq$ means greater than or equal to. The restriction $r \geq R_{\text{Earth}}$ indicates that the equation only applies when r is greater than or equal to the radius of the Earth, $R_{\text{Earth}} = 6.38 \times 10^6\ \text{m}$.

Using Equation 3.33 and the results obtained earlier it is possible to understand a variety of phenomena concerning circular orbits around the Earth such as Example 3.1.

Example 3.1

Many staffed space missions involve low Earth orbits, just beyond the fringes of the Earth's atmosphere. The period of such an orbit is usually about 90 min. Show that this is consistent with what you know about circular motion and the acceleration due to gravity.

Solution

The radius of a low Earth orbit will be only slightly greater than the radius of the Earth, $R_{\text{Earth}} = 6.38 \times 10^6\ \text{m}$ and the magnitude of the acceleration due to gravity in such an orbit will be only slightly less than the corresponding magnitude at the Earth's surface $g = 9.81\ \text{m}\,\text{s}^{-2}$. Adopting g as an approximate value for the centripetal acceleration in a low Earth orbit and using Equation 3.31 to relate that acceleration to the angular speed of a spacecraft close to the surface of the Earth, we have

$$g = R_{\text{Earth}}\omega^2.$$

But we know (from Equation 3.20) that $\omega = 2\pi/T$, where T is the orbital period. It follows that

$$g = \frac{4R_{\text{Earth}}\pi^2}{T^2}$$

and rearranging this to isolate T gives

$$T = 2\pi\sqrt{\frac{R_{\text{Earth}}}{g}}.$$

Substituting the values for R_{Earth} and g we find

$$T = 2\pi\sqrt{\frac{6.38 \times 10^6\ \text{m}}{9.81\ \text{m}\,\text{s}^{-2}}} = 5070\ \text{s} = 84\ \text{min},$$

a reasonable result considering the approximations we have made.

In Section 1 it was stated that satellites in a geostationary orbit were at a distance of about 42 300 km from the centre of the Earth. Using Equation 3.33, it is possible to show that this is indeed the approximate radius for a 24-hour circular Earth orbit. Here is your chance to do just that.

Question 3.10 Find the radius of the geostationary Clarke orbit. ■

3 Simple harmonic motion

3.1 Some examples of simple harmonic motion

This subsection introduces a very important kind of motion that has applications in almost every field of physics.

> Open University students should leave the text at this point and do the multimedia package *Simple harmonic motion*. When you have completed this you should return to the text. The activity will occupy about one hour.

Take a look at the two mechanical systems shown in Figures 3.20 and 3.21. Figure 3.20 shows a weight hanging from a spring of negligible mass. Figure 3.21 shows a heavy body (called a **bob** in this context) suspended from a fixed point by a light inelastic string to form a simple pendulum. In each case, the system is in its **equilibrium position** and the body is not moving.

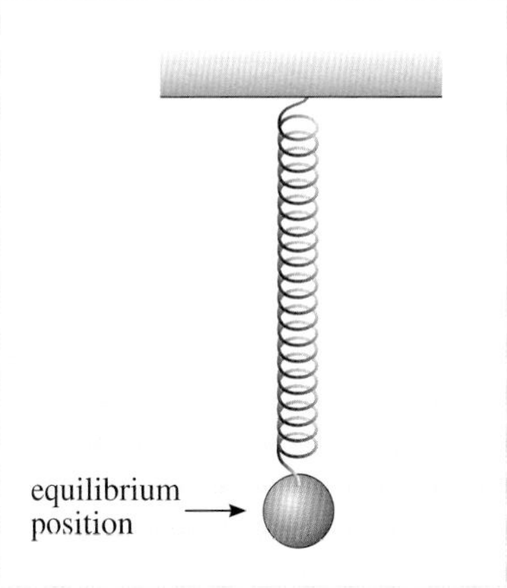

Figure 3.20 Heavy object hanging from a spring.

If you now imagine pulling the weight in Figure 3.20 vertically downwards by a small amount and then releasing it, everyday experience shows that it will oscillate up and down, moving alternately above and below the equilibrium position. Likewise, if the pendulum bob is displaced slightly to one side and then released, it too will oscillate, swinging back and forth around the equilibrium position. In practice these oscillations will eventually cease, but under ideal conditions — in the absence of friction and air resistance — the oscillations would continue forever.

Remarkably, if we limit our attention to such ideal cases, and suppose that the initial displacement is sufficiently small, then it turns out that both the weight on the spring and the bob of the pendulum oscillate in a similar way. In either case, if we introduce an appropriately oriented x-axis, as indicated in Figures 3.22 and 3.23, then we can use the x-coordinate of the oscillating body to measure its displacement from the equilibrium position. A graph showing the instantaneous displacement x plotted against time t will look something like Figure 3.24. The kind of motion described by the figure is a very particular kind of oscillatory motion known as **simple harmonic motion**, a term used so frequently in physics that it is often abbreviated to s.h.m.

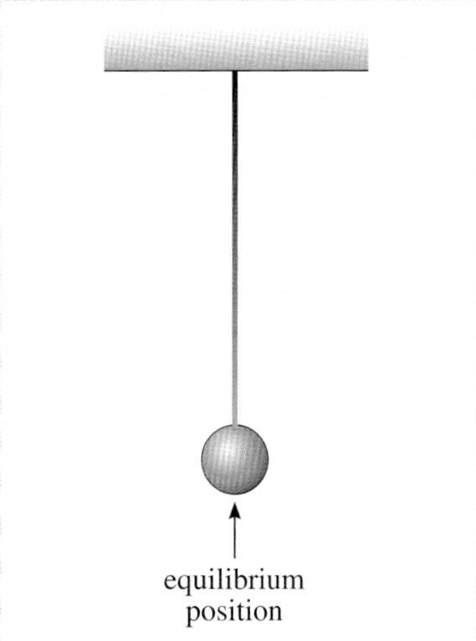

Figure 3.21 Heavy object suspended by an inelastic string.

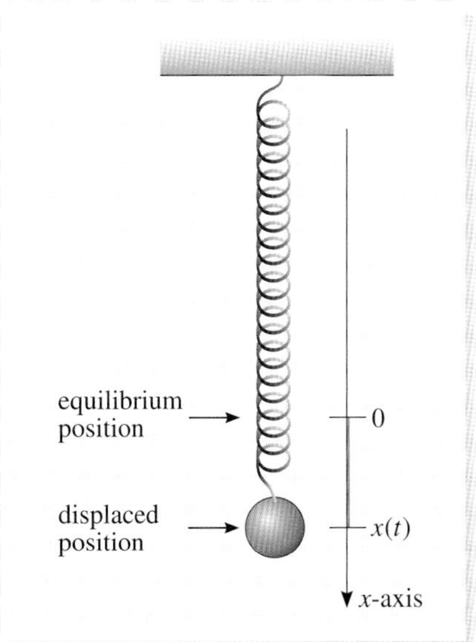

Figure 3.22 Displacement of object in Figure 3.20 from its equilibrium position.

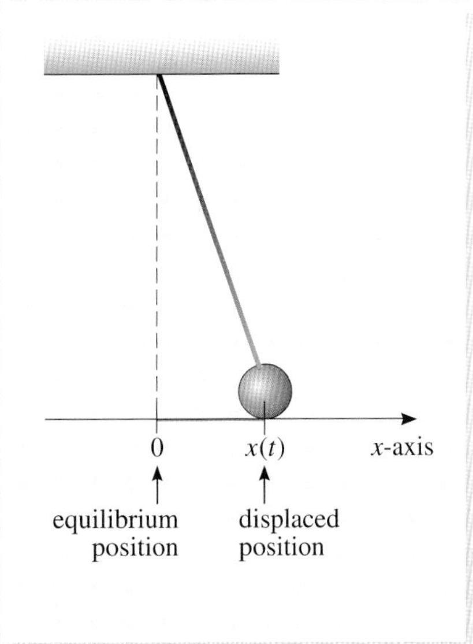

Figure 3.23 Displacement of object in Figure 3.21 from its equilibrium position.

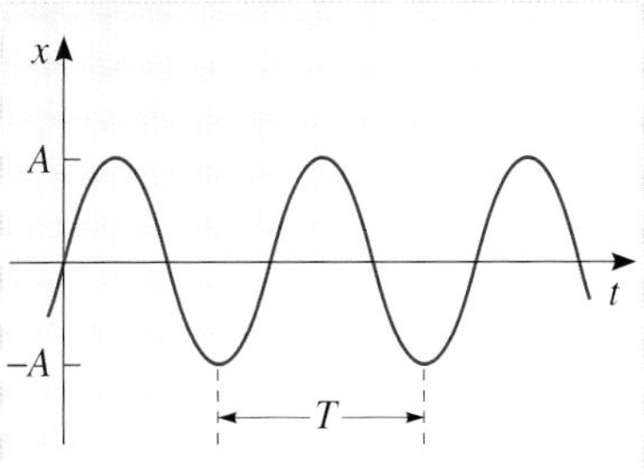

Figure 3.24 Displacement–time graph for the simple harmonic oscillators shown in Figures 3.22 and 3.23. (We have chosen to measure time so that the displacement is zero at time $t = 0$ s.)

As you can see from the graph, two of the quantities that characterize simple harmonic motion are the maximum value A of the displacement x, and the period of the oscillation T. The period is the time taken for one complete oscillation or cycle of the motion, so its reciprocal $1/T$ determines the rate of occurrence of cycles (the number per second) and is known as the **frequency**. If we represent the frequency of the oscillating system by f, then we can define it by the equation

$$f = \frac{1}{T}. \tag{3.34}$$

Frequency could be measured in units of reciprocal seconds (s^{-1}), but there is actually an SI unit of frequency, called the **hertz**, which has the standard abbreviation Hz, and is defined by $1\ Hz = 1\ s^{-1}$. (The reciprocal of a quantity is obtained by dividing 1 by that quantity. Thus the reciprocal of T is $1/T$ or T^{-1}.)

You might well recognize the shape of the displacement–time graph for simple harmonic motion; it is identical to that of a trigonometric function of the kind introduced in Section 2.2. In fact, the displacement from equilibrium of a simple harmonic oscillator may be described generally by the equation

$$x(t) = A\sin(\omega t + \phi). \tag{3.35}$$

Let us examine this equation to see how its features relate to those of Figure 3.24 and to the motion that it describes. The first thing to note about Equation 3.35 is that the $x(t)$ appearing on the left-hand side indicates that x is a function of t; it does *not* represent a product of x and t. In a similar way, the expression $\omega t + \phi$ on the right-hand side of the equation is the *argument* of the sine function. The only *variable* on the right-hand side of Equation 3.35 is the time t. The other quantities A, ω and ϕ, are **parameters** that characterize the motion; their values may vary from one case of s.h.m. to another, but for any particular case they will be constants. Whatever the values of ω and ϕ may be, the value of $\omega t + \phi$ will change with time and the value of $\sin(\omega t + \phi)$ will vary accordingly.

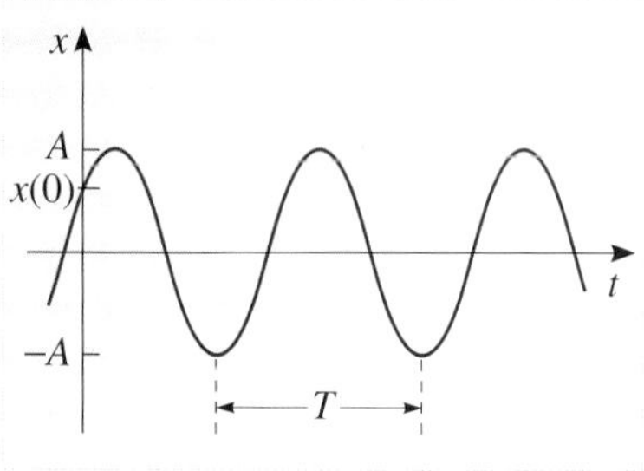

Figure 3.25 The general displacement–time graph for a simple harmonic oscillator. In this case the value of ϕ is not zero and the initial value of the displacement is $x(0) = A\sin(\phi)$.

Comparing $\sin(\omega t + \phi)$ with the general sine function, $\sin(\theta)$, discussed in Section 2.2, you will see that $\sin(\omega t + \phi)$ passes through a full cycle of values every time that its argument $\omega t + \phi$ increases by 2π. It follows that the period of a full cycle of the motion will be $T = 2\pi/\omega$. During a full period, the value of $\sin(\omega t + \phi)$ will vary between +1 and −1; so Equation 3.35 implies that x will vary between A and $-A$. Also, at time $t = 0$ s, the displacement will be $x(0) = A\sin(\phi)$. In drawing Figure 3.24 we have deliberately chosen the time $t = 0$ s in such a way that $x(0) = 0$, implying that $\phi = 0$, but note that this was a choice, not a requirement. Figure 3.25 shows the effect of choosing a different value for ϕ and consequently a different value for $x(0)$. Using these observations we can now define the parameters A, ω and ϕ in Equation 3.35.

The quantity A is known as the **amplitude** of the oscillation. It represents the magnitude of the maximum displacement from the equilibrium position and therefore cannot be negative. It will be measured in the same units as x, usually metres (m).

ω is known as the **angular frequency** of the oscillation. It is related to the frequency f and the period T by the relations

$$\omega = 2\pi f = \frac{2\pi}{T}. \tag{3.36}$$

The angular frequency of s.h.m. is represented by the same symbol ω as the angular speed of circular motion. However the units of angular frequency are s^{-1}, not $rad\ s^{-1}$.

ϕ is called the **phase constant** or the **initial phase** of the oscillation. It represents the value of $\omega t + \phi$ (which is called the **phase**) at time $t = 0$ s, and determines the initial displacement of the oscillator as described by Equation 3.37

$$x(0) = A\sin(\phi). \tag{3.37}$$

Phase is a pure number (like the product ωt), without any units, though it is often treated as an angle and quoted in radians or degrees.

Question 3.11 The period of an oscillator is 0.25 s. What is its frequency?

Question 3.12 The period of an oscillator is 4 s. What is its angular frequency?

Question 3.13 Treating your hand as a particle, write down an approximate mathematical description of its motion during a handshake on the assumption that it is simple harmonic motion. Explain the coordinate system you have chosen to use and include realistic values for the amplitude and angular frequency. Explain the significance of whatever value you have chosen for the initial phase. How faithfully do you think this model represents the actual motion of your handshake? ■

There is a deep link between simple harmonic motion and uniform circular motion (see Figure 3.26). If you look back at Section 2 you will see from Equation 3.18 that uniform circular motion in the anticlockwise sense around a circle of radius R, centred on the origin, occurs when the Cartesian coordinates of the moving particle are given by

$$x(t) = R\cos(\omega t + \theta_0) \tag{3.38}$$

$$y(t) = R\sin(\omega t + \theta_0) \tag{3.39}$$

where θ_0 represents the value of the angular coordinate θ at time $t = 0$ s. As you can see, the y-coordinate (the projection of the circling particle's position onto the y-axis) describes a simple harmonic motion with amplitude R, initial phase $\phi = \theta_0$, and an angular frequency ω that is numerically equal to the angular speed of the circular motion. The relationship becomes even deeper when you realize that the sine and cosine functions are linked by the following general identity

$$\cos(\theta) = \sin(\theta + \pi/2) \quad \text{(for any } \theta\text{, where } \theta \text{ is in radians).} \tag{3.40}$$

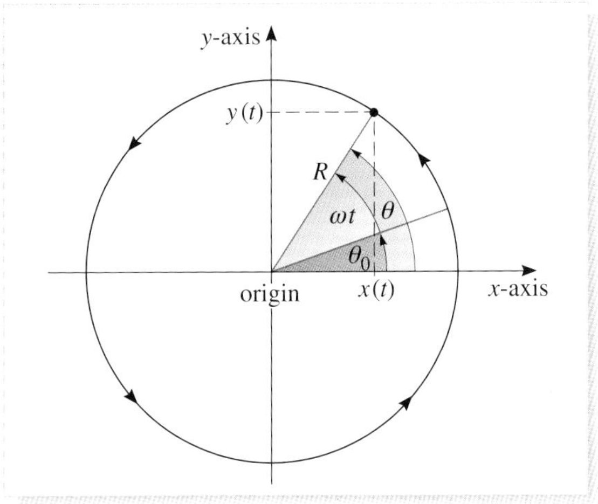

Figure 3.26 Uniform circular motion around a circle of radius R, centred on the origin, in the anticlockwise sense, with angular speed ω. The initial value (at time $t = 0$ s) of the angular coordinate θ is indicated by θ_0.

This identity means that the instantaneous x-coordinate of uniform circular motion (Equation 3.38) may be written as

$$x(t) = R\sin(\omega t + \theta_0 + \pi/2) \qquad (3.41)$$

and this is nothing other than the description of simple harmonic motion with amplitude R, angular frequency ω, and initial phase $\phi = \theta_0 + \pi/2$.

We saw in Chapter 2, that the parabolic motion of a projectile was the result of combining uniform motion in the x-direction with uniformly accelerated motion in the y-direction. What we have now shown is that uniform circular motion is the result of combining a simple harmonic motion in the x-direction with a similar simple harmonic motion in the y-direction that has the same amplitude and angular frequency, but which differs in initial phase by $\pi/2$.

Jules Antoine Lissajous (1822–1880) was a French physicist who used the figures that now bear his name as a means of comparing the characteristics of two oscillating tuning forks.

Combining simple harmonic motions in the x- and y-directions with a variety of amplitudes, angular frequencies and initial phases leads to a range of interesting periodic motions. Some of these, known as Lissajous figures, are shown in Figure 3.27.

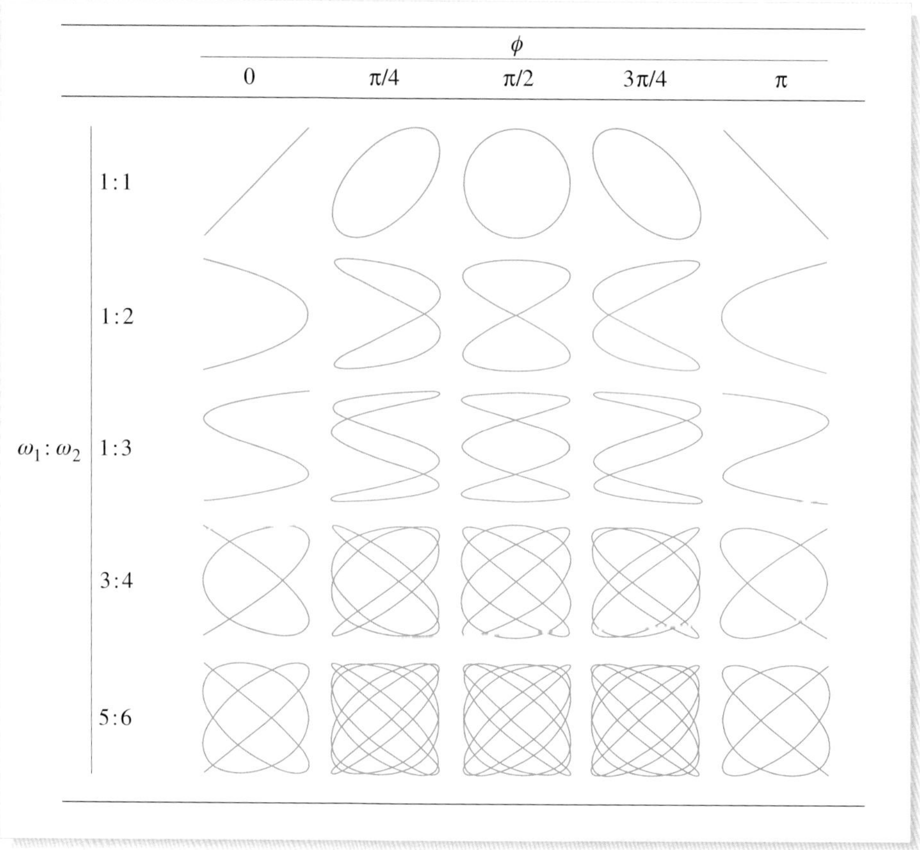

Figure 3.27 Some Lissajous figures that arise from the combination of two perpendicular simple harmonic motions $y(t) = A\sin(\omega_1 t)$ and $x(t) = A\sin(\omega_2 t + \phi)$ where ω_1 and ω_2 are the angular frequencies.

Question 3.14 Express $\sin(\theta + \pi/2)$ and $\sin(\theta + \pi)$ in terms of the cosine function. ■

3.2 Velocity and acceleration in simple harmonic motion

In principle it is a simple matter to work out the velocity and acceleration of a particle executing simple harmonic motion. We already know that the position of such a particle may be described (in a suitably chosen system of coordinates) by the following function of time

$$x(t) = A\sin(\omega t + \phi). \qquad \text{(Eqn 3.35)}$$

We may therefore determine the corresponding velocity v_x at any time t by finding the derivative of $x(t)$, in accordance with Equation 1.14

$$v_x(t) = \frac{\mathrm{d}x}{\mathrm{d}t} = \frac{\mathrm{d}}{\mathrm{d}t}[A\sin(\omega t + \phi)]. \qquad (3.42)$$

Similarly, the acceleration of the particle may be determined by finding the derivative of the velocity, in accordance with Equation 1.15.

$$a_x(t) = \frac{\mathrm{d}v_x}{\mathrm{d}t} = \frac{\mathrm{d}^2x}{\mathrm{d}t^2} = \frac{\mathrm{d}^2}{\mathrm{d}t^2}[A\sin(\omega t + \phi)]. \qquad (3.43)$$

The only drawback with this method is that you need to know how to work out the necessary derivatives which are somewhat more complicated than those we considered in earlier chapters. The results are actually quite simple and are quoted below; their justification (though not a full proof) is considered in Box 3.2.

Box 3.2 Derivatives of the sine and cosine functions

In many applications, it is crucial to know the derivatives of the general sine and cosine functions, $\sin(\theta)$ and $\cos(\theta)$. From the discussion of differentiation in Chapter 1 you should be aware that the derivative of a sine function with respect to its argument θ may be written as $\dfrac{\mathrm{d}\sin(\theta)}{\mathrm{d}\theta}$, and that it too is a function of θ. For any given value of θ, the corresponding value of $\dfrac{\mathrm{d}\sin(\theta)}{\mathrm{d}\theta}$ is given by the gradient of the tangent to the graph of $\sin(\theta)$ at the same value of θ. This is important information, but what would be really useful is a simple expression for the derivative that could be easily evaluated (using a calculator) for any given value of θ. Fortunately such an expression exists, and there is also a similar expression for the derivative of $\cos(\theta)$. In fact

$$\frac{\mathrm{d}\sin(\theta)}{\mathrm{d}\theta} = \cos(\theta) \qquad (3.44)$$

$$\frac{\mathrm{d}\cos(\theta)}{\mathrm{d}\theta} = -\sin(\theta). \qquad (3.45)$$

The following question gives you the chance to justify these claims to some extent.

Question 3.15 Figure 3.28 shows the graphs of (a) $\sin(\theta)$ and (b) $\cos(\theta)$.

(i) From Figure 3.28a, estimate the gradient of $\sin(\theta)$ at $\theta = 0$, $\pi/2$, π and $3\pi/2$. Compare your answers with the values of $\cos(\theta)$ at the same values of π. Are they the same to within the accuracy of your estimates?

(ii) From Figure 3.28b, estimate the gradient of $\cos(\theta)$ at $\pi = 0$, $\pi/2$, π and $3\pi/2$. Compare your results with the values of $\sin(\theta)$ at the same values of θ. How do they differ? ■

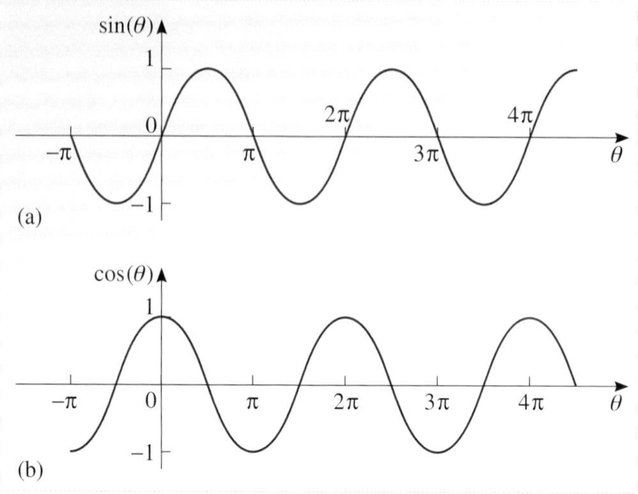

Figure 3.28 The graphs of (a) $\sin(\theta)$ and (b) $\cos(\theta)$.

In many situations (e.g. circular motion or s.h.m.) the argument θ of a trigonometric function will depend on some other quantity such as time. For instance we might have $\theta = \omega t$ where ω is a constant. In such circumstances we often need to know the derivative of $\sin(\omega t)$ or $\cos(\omega t)$ with respect to t rather than with respect to θ. Though we shall not give a proof, the derivatives in this case turn out to be

$$\frac{\mathrm{d}\sin(\omega t)}{\mathrm{d}t} = \omega\cos(\omega t) \tag{3.46}$$

$$\frac{\mathrm{d}\cos(\omega t)}{\mathrm{d}t} = -\omega\sin(\omega t)\cdot \tag{3.47}$$

In the particular case of s.h.m. that is currently of interest, θ is a linear function of t given by $\theta = \omega t + \phi$, where ω and ϕ are both constants. The effect of adding the constant ϕ is to shift the graph of the trigonometric function to the left. As a result the trigonometric function representing the derivative moves to the left by the same amount. In other words

$$\frac{\mathrm{d}\sin(\omega t + \phi)}{\mathrm{d}t} = \omega\cos(\omega t + \phi) \tag{3.48}$$

$$\frac{\mathrm{d}\cos(\omega t + \phi)}{\mathrm{d}t} = -\omega\sin(\omega t + \phi)\cdot \tag{3.49}$$

Question 3.16 It was stated in Chapter 1 that if $f(t)$ is a function of t, and A is a constant, then $\dfrac{\mathrm{d}(Af)}{\mathrm{d}t} = A\dfrac{\mathrm{d}f}{\mathrm{d}t}$. Use this rule together with the above results to determine

$$\frac{\mathrm{d}}{\mathrm{d}t}[A\sin(\omega t + \phi)] \quad \text{and} \quad \frac{\mathrm{d}^2}{\mathrm{d}t^2}[A\sin(\omega t + \phi)]\cdot \quad ■$$

It follows from the results of Question 3.16 that:

If $$x(t) = A\sin(\omega t + \phi) \qquad \text{(Eqn 3.35)}$$

then $$v_x(t) = \frac{\mathrm{d}x}{\mathrm{d}t} = A\omega\cos(\omega t + \phi) \qquad (3.50)$$

and $$a_x(t) = \frac{\mathrm{d}v_x}{\mathrm{d}t} = -A\omega^2\sin(\omega t + \phi)\cdot \qquad (3.51)$$

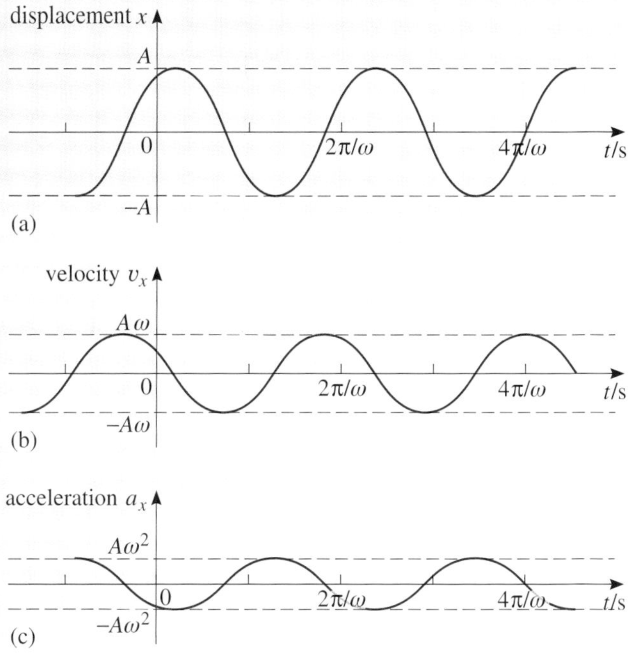

Figure 3.29 Time dependence of (a) the displacement x, (b) the velocity v_x and (c) the acceleration of a particle in simple harmonic motion about $x = 0$.

These are the basic equations descibing simple harmonic motion. The displacement x, the velocity v_x and acceleration a_x are plotted, as functions of time, in Figure 3.29. The graphs (or the equations that describe them) reveal a number of important features about simple harmonic motion:

- All three quantities vary **sinusoidally** with time. That is, each of the three quantities is described by a graph that may be represented by a sine function. In Equation 3.50, the velocity is described by a cosine function, but Equation 3.40 shows that a cosine function *may* be written as a sine function with a suitably chosen initial phase.

- All three quantities are characterized by the same angular frequency ω. They therefore all have the same period, $T = 2\pi/\omega$.
- The maximum value of the displacement is the amplitude A, the maximum velocity is $A\omega$, and the maximum acceleration is $A\omega^2$.
- When the magnitude of the displacement is a maximum (A), the velocity is momentarily zero, and the acceleration has its maximum magnitude of $A\omega^2$. This relationship arises because the velocity reaches its maximum value a quarter of a cycle ahead of the displacement, and the acceleration reaches its maximum a quarter of a cycle ahead of the velocity.

The last of these points may be seen even more clearly if we represent the simple harmonic oscillator's displacement, velocity and acceleration by the following equations, which may be shown to be mathematically equivalent to those given above

$$x(t) = A\sin(\omega t + \phi) \qquad \text{(Eqn 3.35)}$$

$$v_x(t) = A\omega\sin(\omega t + \phi + \pi/2) \qquad (3.52)$$

$$a_x(t) = A\omega^2\sin(\omega t + \phi + \pi). \qquad (3.53)$$

(Note that in Equation 3.35 we distinguish the phase $(\omega t + \phi)$ from the phase constant or initial phase, ϕ. Some authors refer to ϕ as the phase.)

These equations show that when the phase of the displacement is $(\omega t + \phi)$, the phase of the velocity is $(\omega t + \phi + \pi/2)$. This is described by saying that there is a **phase difference** between the displacement and the velocity. The velocity is said to *lead* the displacement by $\pi/2$. Similarly, the acceleration leads the displacement by a phase difference of π.

3.3 The simple harmonic motion equation

From Figure 3.29a and b, or equivalently from Equations 3.35 and 3.51, it can be seen that the displacement $x(t)$ and the acceleration $a_x(t)$ have the same general shape, but they differ in maximum and in sign. In fact, we may write

$$a_x(t) = -\omega^2 x(t) \qquad (3.54)$$

This is sometimes referred to as the **simple harmonic motion equation**. Expressed in words it tells us that

> In simple harmonic motion, the acceleration of the oscillator is proportional to its displacement from the equilibrium position, and the constant of proportionality is the negative quantity $-\omega^2$, where ω is the angular frequency of the motion.

This is a defining characteristic of simple harmonic motion. Any kind of motion in which the acceleration is, at all times, proportional to the displacement, and the constant of proportionality is negative, will be simple harmonic. Moreover, whatever symbol is used to represent the negative proportionality constant, its magnitude will always be equal to the square of the angular frequency and may therefore be used to determine the period and the frequency of the oscillation, using $\omega = 2\pi f = 2\pi/T$.

In practice the s.h.m. equation often appears in the form

$$\frac{d^2x(t)}{dt^2} = -\omega^2 x(t)\cdot \qquad (3.55)$$

An equation like this, which relates a variable quantity such as x to one or more of its derivatives, is called a **differential equation**. Such equations are of fundamental importance throughout physics. The solution to a differential equation is not simply a number or a value, but rather a function, e.g. $x(t)$ in the case of Equation 3.55. The s.h.m. equation is a particularly important differential equation that arises in many areas of physics. It is said to be a **second-order differential equation** because it involves a second derivative (d^2x/dt^2), but no higher derivatives. As you have seen its solution may be written in the form

$$x(t) = A\sin(\omega t + \phi). \qquad \text{(Eqn 3.35)}$$

This is in fact the **general solution** to the s.h.m. equation. This means that *any* solution to the s.h.m. equation may be written in the form of Equation 3.35 provided A and ϕ are chosen appropriately. However, it's worth noting that because of the many equivalent ways of expressing a given trigonometric function, the general solution to the s.h.m. equation may itself be written in a variety of equivalent ways. We have based our discussion on solutions of the form $A\sin(\omega t + \phi)$, but we could equally well have based our whole discussion on solutions of the form

$$x(t) = A\cos(\omega t + \phi) \qquad (3.56)$$

or

$$x(t) = A\sin(\omega t) + B\cos(\omega t). \qquad (3.57)$$

Note that that in the case of Equation 3.57 there is no phase constant as such, but the introduction of a second trigonometric function has also introduced a new constant B. Suitable choices of A and B in this case would enable us to cover the same range of solutions as appropriate choices of A and ϕ in Equations 3.56 or 3.35. Each of these different forms of the solution is quite common, so you should be prepared to meet any of them and to recognize each of them as a description of simple harmonic motion.

3.4 The importance of simple harmonic motion

Simple harmonic motion occurs in musical instruments, electric circuits, loudspeakers, clocks, vibrating machinery, atoms in crystals .... In fact, simple harmonic motion occurs in such a wide range of physical situations, that you may wonder why it is so common. Physically, there are a number of requirements that must be met for simple harmonic motion to occur in a given system.

- The system must have an equilibrium position from which it can be displaced.
- When it is displaced from its equilibrium position it must have a tendency to return to that equilibrium position.
- Whatever the cause of that tendency to return to equilibrium it must produce an acceleration directed towards the equilibrium position that is proportional to the displacement from equilibrium.

There are a number of physical systems that exactly meet these requirements, but there are many more that do so approximately, and almost every vibrating system will meet them provided its displacement from equilibrium is never allowed to become too great. Thus s.h.m. provides a description of many systems, and an approximate description, valid for small displacements, for almost all vibrating systems.

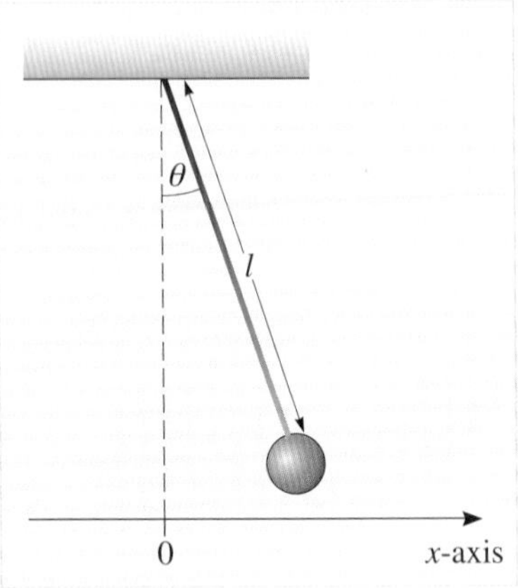

Figure 3.30 A pendulum of length l, with instantaneous angular displacement $\theta(t)$ from its equilibrium position. For small θ we have $x = l\theta$.

Example 3.2

Figure 3.30 shows a pendulum consisting of a bob attached to a light inelastic string of length l, supported from a fixed point. If the pendulum's instantaneous angular displacement from its equilibrium position is denoted by $\theta(t)$, it can be shown that the behaviour of the pendulum is described by the differential equation

$$\frac{d^2\theta(t)}{dt^2} = -\frac{g}{l}\sin[\theta(t)] \tag{3.58}$$

where g is the magnitude of the acceleration due to gravity.

Show that, provided the displacement from equilibrium is sufficiently small, this system can be described by the equation of simple harmonic motion. Find an expression for the period of the pendulum, and hence show that the period is independent of the displacement provided it is small.

Solution

Equation 3.58 does not have the form of the s.h.m. equation, but, for small displacements, we know from Equation 3.26 that $\sin\theta \approx \theta$ provided θ is measured in radians. It follows that for sufficiently small values of θ

$$\frac{d^2\theta(t)}{dt^2} = -\frac{g}{l}\theta(t) \quad (\theta \text{ small and in radians}). \tag{3.59}$$

This is already recognizable as having the form of the s.h.m. equation even though it involves the angular displacement θ. However, if you wish, you may multiply both sides by the constant l, and then use the fact that $l\theta \approx x$ for sufficiently small θ to write

$$\frac{d^2x}{dt^2} = -\frac{g}{l}x. \tag{3.60}$$

This is the s.h.m. equation with the positive constant ω^2 replaced by g/l. It follows that for sufficiently small displacements the pendulum executes s.h.m. with angular frequency

$$\omega = \sqrt{\frac{g}{l}}. \tag{3.61}$$

Since $T = 2\pi/\omega$, the period of the pendulum will be

$$T = 2\pi\sqrt{\frac{l}{g}}. \tag{3.62}$$

This period depends on the length of the pendulum, but not on the displacement, provided it's small enough for the approximations we have used to be valid.

Another reason for the importance of s.h.m. is that many complicated vibrations, such as those caused by earthquakes, which are not themselves simple harmonic, can generally be regarded as a sum of simple harmonic motions. A weight on a spring provides an example of this. A real weight can oscillate up and down, but as Figure 3.31 shows, it can also twist, or swing back and forth in either or both of two independent directions. Such a weight is said to have four **modes of oscillation**. When all four occur simultaneously the resulting motion will be very complicated, but by treating it as a sum of simple harmonic oscillations the analysis can be considerably simplified.

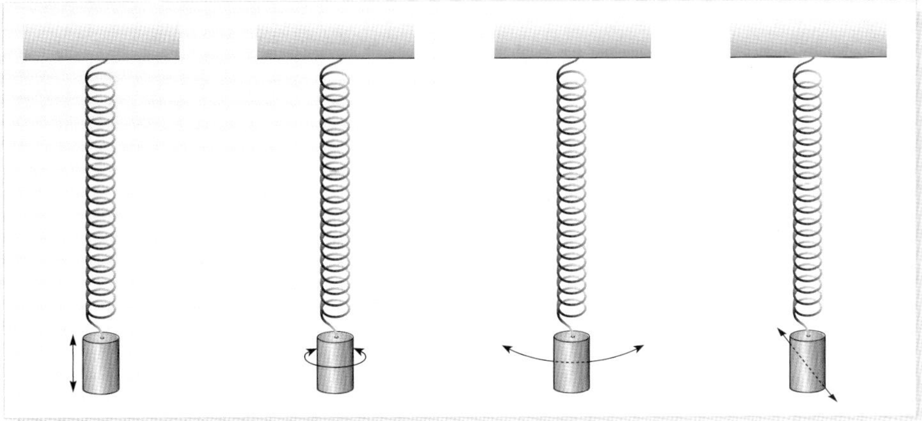

Figure 3.31 The four modes of oscillation of a weight on a spring.

Question 3.17 A cylinder of cross-sectional area A and mass M, floats vertically in calm seawater of density ρ. If its centre is vertically displaced by an amount y from its equilibrium position and then released, it will oscillate up and down. You should assume that the oscillations are described by an equation of the form

$$\frac{d^2y}{dt^2} = -\frac{A\rho g}{M}y\,. \qquad (3.63)$$

ρ is the Greek lower case letter rho.

What will be the period of those oscillations for a partly-filled oil drum, floating in this way, if its diameter is 0.7 m, its mass is 80 kg and the density of seawater is $1.025 \times 10^3\,\text{kg m}^{-3}$? ■

4 Orbital motion

4.1 A note on the ellipse

In Section 2 we considered circular satellite orbits. However, observation shows that, more generally, the orbits may be elliptical. An **ellipse**, like a circle or a parabola, is a member of the family of curves known as conic sections. In terms of a two-dimensional Cartesian coordinate system, an ellipse may be defined by choosing two lengths a and b, and then plotting all the points (x, y) that satisfy the equation

$$\frac{x^2}{a^2} + \frac{y^2}{b^2} = 1 \quad (a \text{ greater than or equal to } b) \qquad (3.64)$$

as shown in Figure 3.32. Such an ellipse is said to have a **semimajor axis** of length a, and a **semiminor axis** of length b.

Figure 3.32 shows that an ellipse looks a bit like a partially squashed circle. A measure of how much the ellipse differs from a circle is given by its **eccentricity** e, which has the definition

$$e = \frac{1}{a}\sqrt{a^2 - b^2} \quad (a \text{ greater than or equal to } b). \tag{3.65}$$

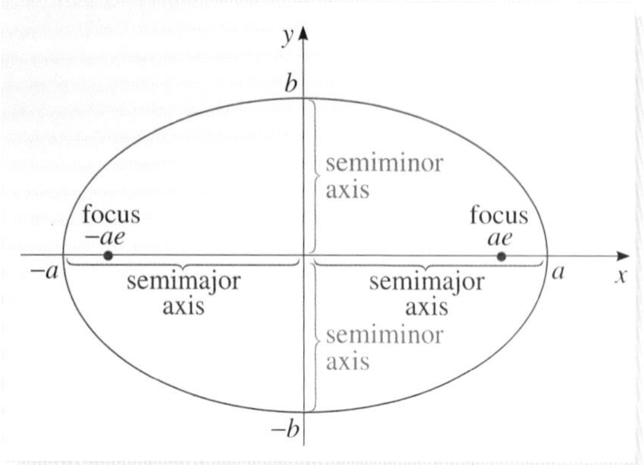

Figure 3.32 General features of an ellipse.

Figure 3.33 shows a number of ellipses with different values of e (but the same value for a). The smaller the eccentricity, the more circular the ellipse. A circle is actually a special case of an ellipse with $a = b$, and therefore $e = 0$.

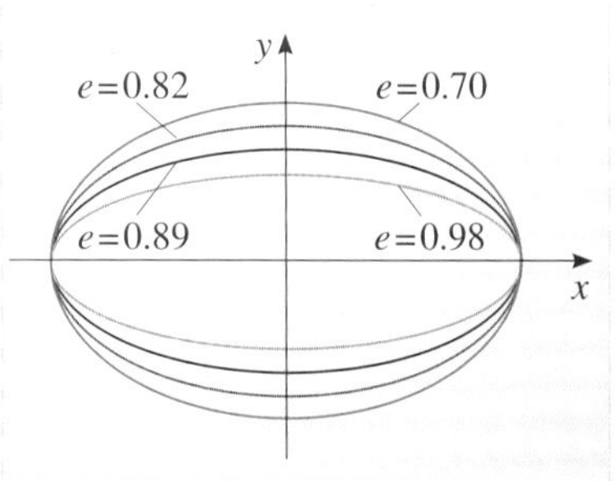

Figure 3.33 Ellipses with the same semimajor axis a, but different values of the eccentricity, e.

An ellipse has two special points on its major axes, each called a **focus** (plural *foci*). These are located at the points $(ae, 0)$ and $(-ae, 0)$. One of the things that makes them special is that the sum of the two distances from any point on the ellipse to the two foci is a constant, equal to $2a$, as shown in Figure 3.34. In fact, one way of drawing an ellipse of semimajor axis a and eccentricity e is to fix the ends of a piece of string of length $2a$ at the foci of the ellipse (a distance $2ae$ apart). Then, while using a pencil to keep the string taut, move that pencil around the foci so as to trace out a closed curve. That curve will be the required ellipse. Try it!

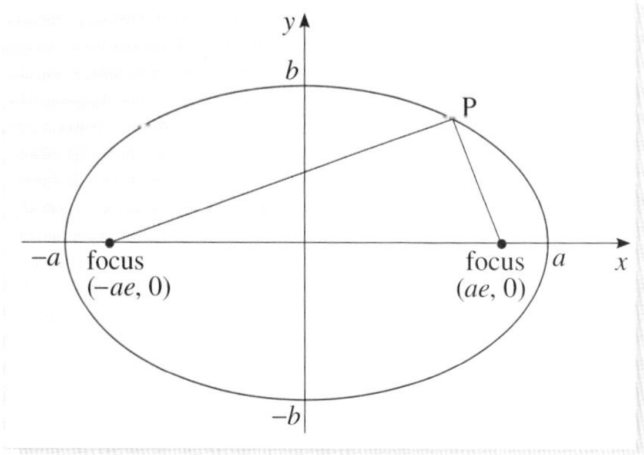

Figure 3.34 For each point on the ellipse the sum of the distances to the two foci is constant and equal to $2a$.

Question 3.18 A particle follows a path described by the equation $x^2 + 4y^2 = 16\,\text{m}^2$. Show that the path is an ellipse and find, in the notation of Equation 3.65, the values of a, b and e for this particular ellipse. Also find the positions of the foci.

Question 3.19 Using Equation 3.65, show that the eccentricity of any ellipse must satisfy the relation $0 \le e < 1$. Describe the curves that are obtained as e approaches its limiting values of 0 and 1. (The symbol $\le$ means 'less than or equal to', the $<$ symbol just means 'less than.' Similar symbols, $>$ and $\ge$, indicate 'greater than' and 'greater than or equal to', respectively.) ■

4.2 Kepler's laws

The planets (Figure 3.35) have always fascinated scientists, and the struggle to understand the nature of planetary orbits has played an important part in the development of physics. Even before telescopes began to be used in astronomy, quite precise measurements had been made of planetary movements in an effort to determine the size, shape and disposition of each planet's orbit. The most accurate of these early measurements were made by the Danish astronomer Tycho Brahe (Figure 3.36a), the last of the great naked-eye astronomers.

Figure 3.35 A montage of the planets.

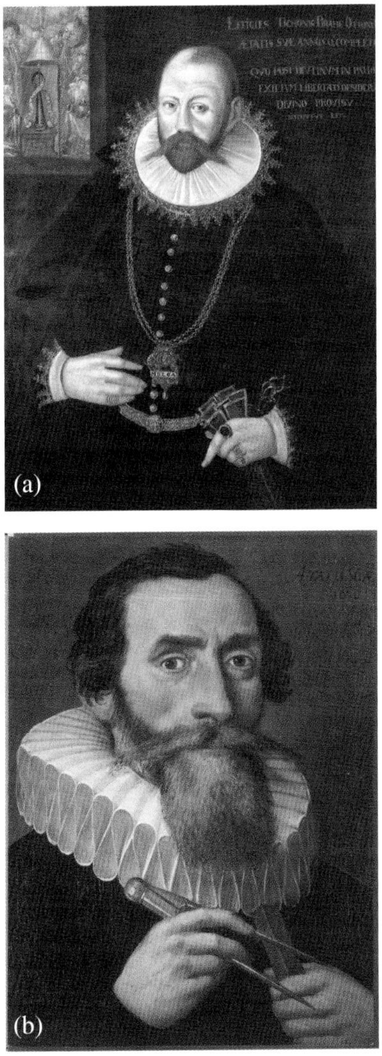

Figure 3.36 (a) Tycho Brahe (1546–1601), (b) Johannes Kepler (1571–1630).

Following Brahe's death, his observational data was used by his former assistant, Johannes Kepler (1571–1630), to determine the nature of the planetary orbits. The task was a difficult one, but Kepler (Figure 3.36b) succeeded and was eventually able to summarize his main results in three descriptive laws. These are now known as **Kepler's laws**. With hindsight, we now know that these laws are not entirely accurate. Gravitational interactions between planets cause them to depart slightly from the behaviour described by Kepler's laws. Nonetheless, these laws encapsulate an essential understanding of planetary orbits and continue to be of great value.

> **Kepler's first law**
> The orbit of each planet in the Solar System is an ellipse with the Sun at one focus.

This law is illustrated in Figure 3.37.

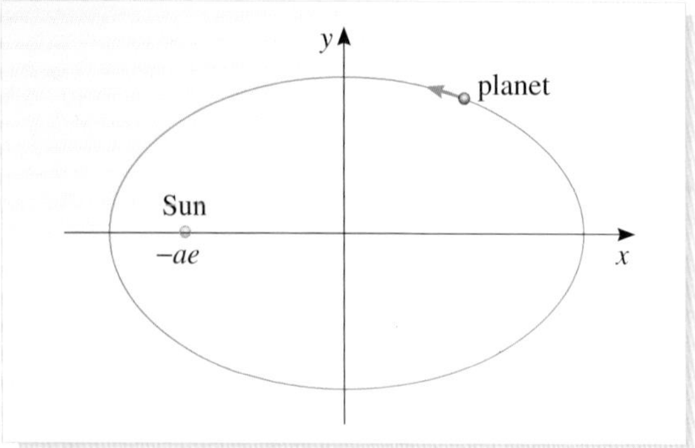

Figure 3.37 An illustration of Kepler's first law.

Kepler's second law
A radial line from the Sun to a planet sweeps out equal areas in equal intervals of time.

The second law is illustrated in Figure 3.38. If the areas A_1 and A_2 are equal, then the time taken to traverse the corresponding parts of the orbit will also be equal. However, the arcs bounding these areas are of different lengths so a planet must move across the longer arc at a higher speed. The second law therefore implies that a planet must move more quickly as it gets closer to the Sun.

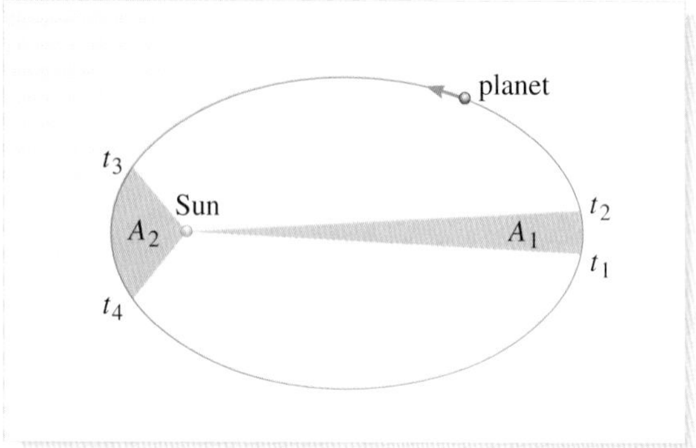

Figure 3.38 An illustration of Kepler's second law. Equal areas ($A_1 = A_2$) are swept out in equal times ($t_2 - t_1 = t_4 - t_3$), so the planet moves faster when closer to the Sun.

Kepler's third law
The square of the orbital period of each planet is proportional to the cube of its semimajor axis.

The period of a planetary orbit is the time taken for a planet to complete one full circuit of the Sun. In the case of the Earth it is one year. Kepler's third law tells us that there is a single constant, let's call it K, such that for any body that orbits the Sun with period T and semimajor axis a

$$\frac{T^2}{a^3} = K. \tag{3.66}$$

Kepler's third law is illustrated in Figure 3.39 for three hypothetical planets with periods T_1, T_2, T_3, and semimajor axes a_1, a_2, a_3. Since the constant K in Equation 3.66 is the same for all three bodies, we can write

$$\frac{T_1^2}{a_1^3} = \frac{T_2^2}{a_2^3} = \frac{T_3^2}{a_3^3}. \tag{3.67}$$

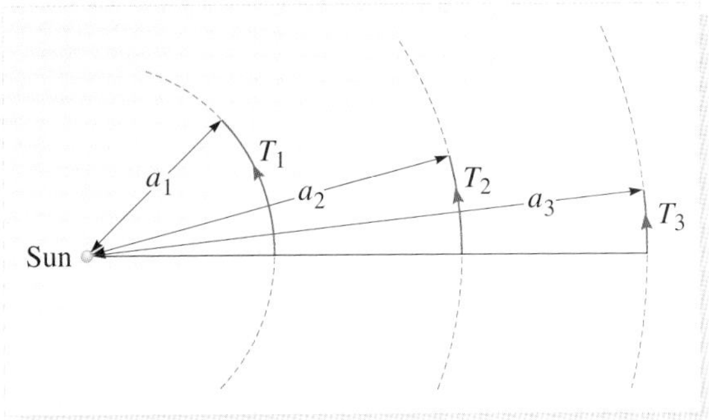

Figure 3.39 An illustration of Kepler's third law. Three hypothetical planets have periods T_1, T_2, T_3, and semimajor axes a_1, a_2, a_3. In each case T^2/a^3 will have the same constant value. This means that $T \propto a^{3/2}$ and, in the case of circular orbits, implies that the orbital speed decreases as the square root of the distance from the Sun increases, so the distance travelled in a fixed time also decreases.

Table 3.2 gives the semimajor axes and periods for the planets. Using these data it is easy to confirm Kepler's third law. This is the subject of Example 3.3.

Table 3.2 Orbital data for the planets.

Planet	Semimajor axis, $a/10^6$ km	Orbital period, T/years
Mercury	57.87	0.241
Venus	108.1	0.615
Earth	149.5	1.00
Mars	227.8	1.881
Jupiter	777.8	11.86
Saturn	1423	29.46
Uranus	2870	84.01
Neptune	4510	164.8
Pluto	5944	247.7

Example 3.3

Using the data given in Table 3.2, plot a graph of T against $a^{3/2}$. Does your result confirm Kepler's third law?

Solution

The required graph is shown in Figure 3.40. As can be seen, to a good approximation, the graph is a straight line. This shows that

$$T \propto a^{3/2}.$$

If these quantities are proportional then their squares will also be proportional

i.e. $$T^2 \propto a^3.$$

This confirms Kepler's third law. (The reasons for plotting T against $a^{3/2}$ rather than T^2 against a^3 are that (i) fewer calculations are needed to obtain the plotted quantities, and (ii) the plotted points are more evenly spread out, which makes for a more accurate graph.)

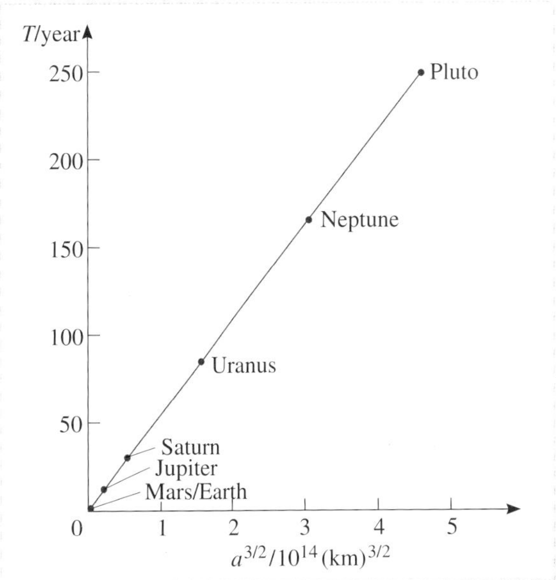

Figure 3.40 Graph of T against $a^{3/2}$ for the planets. On the scale of the graph it is difficult to plot separate points for Earth and Mars. The points for Venus and Mercury cannot be shown at all.

Our understanding of planetary motions has greatly increased since Kepler's time, thanks largely to the work of Newton and his successors. As a result, we now know that the constant K in Equation 3.66 is mainly determined by the mass of the Sun. As a result, we expect Kepler's laws to apply to other systems of orbiting bodies, though the parameter K will change from one system to another.We shall return to this subject in *Predicting motion*, the next book in the series.

Question 3.20 In 1997 the first planet to orbit a normal star other than the Sun was discovered. This planet, which orbits the star 51 Pegasi, has an orbital period of 4.23 days and a semimajor axis of about 7.50×10^6 km. If the system of 51 Pegasi also contained another planet orbiting with the same semimajor axis as that of the Earth's orbit, what would be the orbital period of that planet? ■

4.3 Position, velocity and acceleration in orbital motion

When describing s.h.m. and uniform circular motion we were able to write down simple expressions for the position and velocity as functions of time. In the case of orbital motion however, such simple expressions do not exist. This is true even when interactions between planets are ignored and Kepler's laws are treated as exact. Certainly, there are methods that make it possible to predict where an orbiting body will be at any particular time, and what its velocity will be, but those methods are too complicated to treat here. Nonetheless, there are some important points about orbital motion that should be appreciated, and which do not require advanced mathematics for their description.

In what follows we will restrict our attention to a system consisting of a single satellite orbiting an isolated planet. We shall suppose Kepler's laws to be exact for this system, so that the satellite moves in an ellipse with the planet at one focus, sweeping out equal areas in equal times, and with an orbital period that is related to the orbit's semimajor axis by $T^2 = Ka^3$, for some particular value of K. Under these circumstances:

1. A determination of the position and velocity of the satellite at any instant will uniquely determine its orbit. Since modern radar equipment can provide information about the direction, distance and velocity of an object simultaneously, it follows that a satellite's orbit may be determined very quickly from radar observations.
2. In the absence of any information about velocity, the orbit of the satellite may be determined by measuring its position vector at two different points on the orbit and the time it takes to move between those points. The necessary position data may be obtained from radar observations. The fact that we need to know two positions *and* the time that separates them indicates that more than one orbit passes through two given points; the time is needed to determine which of the possible orbits is the correct one.
3. In the absence of information about velocities and distances, it is still possible to uniquely determine a satellite's orbit from three observations of its direction, together with accurate timings. The first person to develop a method of doing this was Sir Isaac Newton in his *Principia*. His method was soon applied by Edmond Halley, who used it to determine the orbit of the comet that now bears his name (Figure 3.41). The problem is still an important one since astronomers often need to calculate an orbit from just a few visual sightings.
4. Despite the complicated behaviour of position and velocity, the acceleration of the satellite will always be directed towards the focus occupied by the planet, and will obey an inverse square law of the kind discussed earlier, in the context of circular orbits. You will discover the reason for this in *Predicting motion*.

Figure 3.41 Halley's comet.

4.4 Transfer orbits and interplanetary flights

Many space vehicles have already been sent to other planets, and more missions are planned for the future.

There are clearly many factors involved in the planning of interplanetary flights, but one aim is to reduce the total amount of fuel required. It turns out that the most fuel-efficient way of getting directly from the Earth to another planet is to follow what is known as a **Hohmann transfer orbit**. For missions to the more distant planets it can be even more efficient to travel indirectly, closely approaching another planet along the way and picking up a gravitational boost from it, but we shall not consider that here.

A Hohmann transfer between the Earth and Mars is shown in Figure 3.42, where we have assumed that the orbits of the two planets are circular and **coplanar**. (Coplanar orbits are ones that lie in the same plane.) The Hohmann transfer orbit is part of an ellipse with the Sun at one focus and with the orbits of the two planets just touching the ellipse. The space vehicle leaves the Earth at a tangent to its orbit and arrives at Mars tangentially to its orbit. Example 3.4 concerns such a transfer orbit.

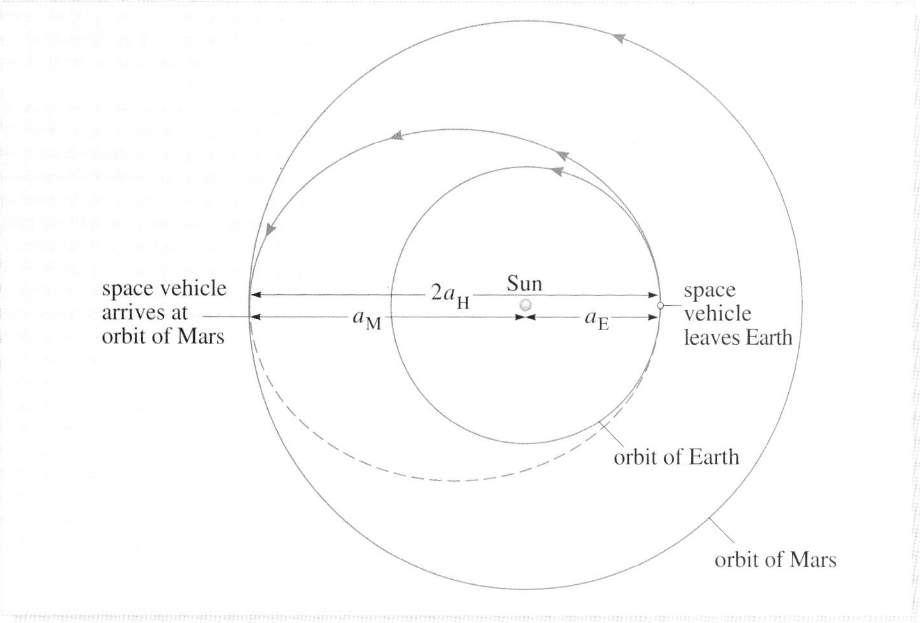

Figure 3.42 Hohmann transfer orbit between Earth and Mars as discussed in Example 3.4. You can see that $2a_H = a_M + a_E$.

Example 3.4

Use Kepler's laws to calculate the time required for a Hohmann transfer from the Earth to Mars.

Solution

The path of the space vehicle is part of an elliptical orbit about the Sun. While following the transfer, the space vehicle will therefore obey Kepler's laws, just like the planets. Using the subscripts H, E and M to denote the Hohmann transfer orbit, the Earth and Mars orbits, respectively, we can write

$$\frac{T_\mathrm{H}^2}{a_\mathrm{H}^3} = K \tag{3.68}$$

$$\frac{T_\mathrm{E}^2}{a_\mathrm{E}^3} = K \tag{3.69}$$

$$\frac{T_\mathrm{M}^2}{a_\mathrm{M}^3} = K \tag{3.70}$$

where T_H, T_E and T_M are the orbital periods and a_H, a_E and a_M are the semimajor axes where K is a constant. It follows that

$$a_\mathrm{H} = \left(\frac{T_\mathrm{H}^2}{K}\right)^{1/3} \tag{3.71}$$

$$a_\mathrm{E} = \left(\frac{T_\mathrm{E}^2}{K}\right)^{1/3} \tag{3.72}$$

$$a_\mathrm{M} = \left(\frac{T_\mathrm{M}^2}{K}\right)^{1/3}. \tag{3.73}$$

Now, from Figure 3.42 we also know that

$$2a_\mathrm{H} = a_\mathrm{E} + a_\mathrm{M}. \tag{3.74}$$

Substitution of Equations 3.71 to 3.73 into Equation 3.74 gives

$$2\left(\frac{T_\mathrm{H}^2}{K}\right)^{1/3} = \left(\frac{T_\mathrm{E}^2}{K}\right)^{1/3} + \left(\frac{T_\mathrm{M}^2}{K}\right)^{1/3} \tag{3.75}$$

which can be simplified to give

$$T_\mathrm{H} = \frac{T_\mathrm{E}}{(2)^{3/2}}\left[1 + \left(\frac{T_\mathrm{M}}{T_\mathrm{E}}\right)^{2/3}\right]^{3/2}. \tag{3.76}$$

Since the approximate orbital periods are $T_\mathrm{E} = 1.00$ year and $T_\mathrm{M} = 1.88$ years, Equation 3.76 gives $T_\mathrm{H} = 1.42$ years. However, this is not the time taken to get to Mars because the space vehicle only makes one-half of an orbit. The time of flight is therefore $T_\mathrm{H}/2 = 7.09 \times 10^{-1}$ years, which is about 259 days. This is similar to the duration of actual missions to Mars.

Question 3.21 Using a Hohmann transfer orbit, calculates how long a journey would take from the Earth to Saturn. You can take the period of Saturn's (approximately circular) orbit to be 29.5 years. ■

In this book we have described many different kinds of motion, but we have not explained why they occur. For example, why does planetary motion obey Kepler's laws so well? To answer questions like this, we need a dynamical theory of motion, that is we need to understand the effects of forces and the role played by the mass of the body. The study of *dynamics* is the subject of *Predicting motion*.

5 Closing items

5.1 Chapter summary

1. Periodic motion is repetitive. Its period is the shortest time T such that $\boldsymbol{r}(t+T) = \boldsymbol{r}(t)$, for all values of t.
2. The radian is a unit of angle, usually abbreviated as rad and defined by 2π rad = 360°. Consequently, 1 rad = 57.296° and the length of a circular arc of radius R that subtends an angle α at the centre of the circle is $s_{\text{arc}} = \alpha R$, provided α is measured in radians.
3. A circle of radius R, centred on the origin may be described, in Cartesian coordinates, by the equation
$$x^2 + y^2 = R^2. \qquad (3.5)$$
The position vector of any point on such a circle can be written
$$\boldsymbol{r} = (x, y) = (R\cos\theta, R\sin\theta), \qquad (3.6)$$
where $\sin\theta$ and $\cos\theta$ are trigonometric functions, defined for all values of θ, and possessing the properties that $\sin(\theta + 2\pi) = \sin\theta$, $\cos(\theta + 2\pi) = \cos\theta$, $\sin(-\theta) = -\sin\theta$ and $\cos(-\theta) = \cos\theta$.
4. For a particle in uniform circular motion of radius R, about the origin, the position at any time t may be written
$$\boldsymbol{r} = (x, y) = (R\cos(\pm\omega t + \theta_0), R\sin(\pm\omega t + \theta_0)), \qquad (3.18)$$
where the + sign indicates anticlockwise motion, the − sign indicates clockwise motion, θ_0 is the angular position at $t = 0$ s and ω represents the (constant) angular speed of the motion, defined by
$$\omega = \left|\frac{d\theta}{dt}\right|. \qquad (3.14)$$
5. For a particle in uniform circular motion with angular speed ω the period is $T = 2\pi/\omega$. The instantaneous velocity is tangential to the circle and has magnitude
$$v = r\omega. \qquad (3.22)$$
The centripetal acceleration is directed towards the centre of the circle and has magnitude
$$a = v\omega = r\omega^2 = v^2/r. \qquad (3.32)$$
6. For a particle in simple harmonic motion about its equilibrium position $x = 0$, the displacement, velocity and acceleration at time t are:
$$x(t) = A\sin(\omega t + \phi) \qquad (3.35)$$
$$v_x(t) = A\omega\cos(\omega t + \phi) \qquad (3.50)$$
$$a_x(t) = -A\omega^2\sin(\omega t + \phi) \qquad (3.51)$$

where A is the amplitude, ω is the angular frequency, and ϕ is the initial phase, i.e. the value of the phase $(\omega t + \phi)$ at $t = 0$ s. Note that the initial displacement of the oscillator will be $x(0) = A \sin \phi$, and that its frequency will be $f = 1/T$. The frequency f is measured in units of hertz (Hz), where 1 Hz = 1 s^{-1}. The angular frequency is measured in s^{-1} and $\omega = 2\pi f$.

7 The equation of simple harmonic motion is

$$\frac{d^2x(t)}{dt^2} = -\omega^2 x(t). \tag{3.55}$$

This is a second-order differential equation with the general solution $x(t) = A \sin(\omega t + \phi)$ where A and ϕ are constants. It shows that a characteristic feature of s.h.m. is that the acceleration is always directed towards the equilibrium position and is proportional to the displacement from that position, i.e. $a_x = -\omega^2 x$.

8 Simple harmonic motion is widespread in nature: examples include the oscillations of springs and pendulums and almost every other kind of oscillator, provided the amplitude is sufficiently small. Even circular motion can be treated as a combination of perpendicular simple harmonic motions with identical amplitudes and angular frequencies, and a phase difference of $\pi/2$.

9 An ellipse of semimajor axis a and semiminor axis b may be described by the equation

$$\frac{x^2}{a^2} + \frac{y^2}{b^2} = 1 \quad (\text{with } a \geq b). \tag{3.64}$$

Such an ellipse has eccentricity $e = \frac{1}{a}\sqrt{a^2 - b^2}$, where $0 \leq e < 1$, and contains two foci, located at the points $(ae, 0)$ and $(-ae, 0)$. The sum of the two distances from any point on the ellipse to the two foci is a constant, equal to $2a$.

10 Kepler's laws describing the motion of the planets about the Sun are:

(i) The orbit of each planet in the Solar System is an ellipse with the Sun at one focus.

(ii) A radial line from the Sun to a planet sweeps out equal areas in equal intervals of time.

(iii) The square of the orbital period of each planet is proportional to the cube of its semimajor axis.

5.2 Achievements

Now that you have completed this chapter, you should be able to:

A1 Understand the meanings of all the newly defined (emboldened) terms introduced in the chapter.

A2 Define the position of a given point in a plane in terms of Cartesian coordinates (x, y) and plane polar coordinates $[r, \theta]$.

A3 Use the radian as a unit of angular measure and convert between radians and degrees. Make use of radians in the calculation of arc lengths.

A4 Describe uniform circular motion, and use the concepts of period, speed, angular speed and centripetal acceleration to solve a variety of problems relating to that motion.

A5 Recall and use the more common properties of the sine and cosine functions, and use their inverse functions (arcsine and arccosine) when appropriate.

A6 Describe simple harmonic motion, and use the concepts of amplitude, frequency, period, angular frequency, phase constant and phase difference to solve a variety of problems relating to s.h.m.

A7 Recognize and write down the simple harmonic motion equation, and explain its physical significance.

A8 Define an ellipse and describe its common properties.

A9 Recall and use Kepler's laws.

A10 Describe various features of orbital motion, as they relate to planetary satellites, both natural and artificial.

5.3 End-of-chapter questions

Question 3.22 The Moon orbits the Earth once every 27 days and 7 hours. If the angular position of the Moon as measured from the centre of the Earth is θ, calculate the change $\Delta\theta$ in the value of θ corresponding to an interval Δt of 2 days and 10 hours. You should assume that the orbit is circular, and give your answer in both radians and degrees.

Question 3.23 (a) Venus orbits the Sun in an approximately circular orbit once every 0.615 years. Calculate the angular speed of Venus. (b) The radius of the orbit of Venus about the Sun is 1.08×10^8 km. Calculate the centripetal acceleration of Venus.

Question 3.24 (a) Sketch graphs of $\sin\theta$ and $\cos\theta$. (b) From your graph of $\cos\theta$, you should be able to see that $\cos\theta = 1$ for $\theta = 2\pi n$, where n is any integer (i.e. any whole number). Write down similar general expressions in terms of n for the solutions to: (i) $\sin\theta = 0$, (ii) $\cos\theta = 0$, (iii) $\sin\theta = 1$, (iv) $\cos\theta = -1$.

Question 3.25 Show that $x(t) = C\cos(\omega t + \phi)$, where C, ω and ϕ are constants, is a solution of the differential equation $d^2x/dt^2 = -\omega^2 x(t)$.

Question 3.26 A particle executes simple harmonic motion with a period of 0.541 s. (a) Calculate the frequency and angular frequency of the particle.

(b) The amplitude of the motion is 3.52 m and the displacement of the particle from its equilibrium position is 2.34 m at time $t = 0$ s. Find a general expression for the displacement of the particle from its equilibrium position at any time t.

Question 3.27 Show that the equation $x^2 + 9y^2 = 9\,\text{m}^2$ represents an ellipse. Find the values of the semimajor and semiminor axes, the foci and the eccentricity.

Question 3.28 A hypothetical planet follows an elliptical orbit with the Sun at one focus. The minimum and maximum distances from the Sun are 1.0×10^8 km and 1.0×10^{10} km, respectively. What are the lengths of the semimajor and semiminor axes, a and b of the planet's orbit?

Question 3.29 Astronomers in the year 2150 claim to have found five planets orbiting a nearby star. They also claim that the planetary orbits are ellipses, with the periods and semimajor axes given in Table 3.3. By plotting a suitable graph, test whether or not these results are consistent with Kepler's third law. ■

Table 3.3 Orbital data for Question 3.29.

Semimajor axis $a/10^6$ km	Orbital period T/years
50.0	0.200
125	1.25
172	2.37
221	3.91
287	6.59

Chapter 4 Consolidation and skills development

1 Introduction

This final chapter has two main aims. Firstly, we introduce a computer package called *Physica* which will provide you with a resource of information and tutorial support that you can use throughout the course. Secondly, we aim to help you to consolidate what you have learned from this book.

To set the scene, we start off in Section 2 with a brief overview of Chapters 1 to 3 emphasizing the fundamental physical concepts and their relevance to other areas of physics. This is followed, in Section 3, by the introduction to *Physica* with a set of exercises designed to help you get to know the package and see how it can help you solve problems and access information.

The remaining sections are devoted to consolidation, starting in Section 4 with a review of a variety of skills you should have acquired from this book. Then Section 5 provides a short-answer test of basic skills and knowledge, followed in Section 6 by a set of interactive questions testing your ability to use your knowledge and skills to solve problems. Finally, Section 7 directs you to some longer problems that you can try with the aid of *Physica*.

2 Overview of Chapters 1 to 3

In the first three chapters of this book you have seen how to describe the motion of a particle. In the next book, *Predicting motion*, you will see how this motion can be explained in terms of the mass of the particle and the forces acting it. Later books move onto other topics.However, you will see that the underlying concepts in this book reappear in other physical contexts in those later books. It is this underlying conceptual unity that makes physics a single subject despite the wide range of topics you see in the titles of the eight books.

In Chapter 1 we introduced motion in one dimension, i.e. the motion of a particle along a straight line. Here we introduced the ideas of a coordinate axis and an origin so that the position x of a particle can be specified. Then we identified the fundamental variables of motion, displacement s_x, velocity v_x and acceleration a_x, and found it useful to display the motion graphically using displacement–time graphs and velocity–time graphs.

A key concept is the slope or gradient of a straight line and, more generally, the gradient at any point on a curve. Velocity and acceleration, and many other quantities in physics, are specified in terms of gradients.

The need for precision in our terminology and definitions led to the introduction of some mathematics.

The concept of a function $f(y)$ was introduced to give precision to the description of how one variable f depends on another variable y. Linear functions, $f(y) = Ay + B$, and quadratic functions, $f(y) = Ay^2 + By + C$, are of particular importance since they occur in many different areas of physics.

A mathematical operation on functions called differentiation was introduced to describe the rate of change of a variable f with respect to another variable y. The result of differentiating a function $f(y)$ is the derived function or derivative, denoted by $\mathrm{d}f/\mathrm{d}y$. The derivative is a precise expression for the gradient at any point on the graph of the function. This gives precision to the definitions of velocity and acceleration. Derivatives will be used throughout this course whenever we need to describe rates of change.

The most important type of motion considered in Chapter 1 was motion with uniform (i.e. constant) acceleration. Examples considered were uniformly accelerated vehicles and objects falling freely in drop-shafts. You will see in the next book that acceleration is the key concept in understanding the effects of forces on a particle. This is why we devote so much time to the concept of acceleration in this book.

In Chapter 2 we extended the study of motion from one dimension to two dimensions, that is, we considered the motion of a particle along a curved path in a plane. Again, the important physical variables are position, displacement, velocity and acceleration, but here these variables are vectors, i.e. they are quantities having direction in space as well as magnitude, and it becomes essential to use vector notation (i.e. bold printing or underlining of symbols) to distinguish vector quantities from scalars such as time.

You saw that vectors of the same kind can be combined with one another by a kind of addition that is described geometrically by the triangle rule.

We can also multiply a vector by a number λ or by any other scalar.

Using a two-dimensional Cartesian coordinate system, any vector $\boldsymbol{a}$ in the xy-plane can be resolved (or projected) into its two components a_x and a_y, hence the ordered pair representation of a vector, $\boldsymbol{a} = (a_x, a_y)$.

This idea of resolving vectors into orthogonal (i.e. perpendicular) components is immensely important. It means that we can describe any motion in a plane simply as the combined effect of two straight-line motions along the x- and y-directions.

Similarly in three dimensions, we set up a right-handed coordinate system, and any motion along a curved path in space is simply the combined effect of three component motions along the x-, y- and z-axes.

As you work through this course you will see that many other physical quantities, such as force, momentum and electric field, are also vectors. In almost all problems involving vector quantities we first set up a suitable two- or three-dimensional coordinate system and then resolve the vectors into their components. In this way we can often express a complex three-dimensional problem as two or three simpler one-dimensional ones.

The vector problem considered in detail in Chapter 2 is projectile motion. Here the motion is in a vertical plane and the vectors are resolved into vertical and horizontal components. The motion is then described as a combination of uniform acceleration under gravity in the vertical direction and motion with uniform velocity in the horizontal direction, with a common time of flight.

Chapter 3 introduced you to examples of periodic motion, starting with the case of an Earth satellite moving in an almost circular orbit with nearly constant speed.

Basic parameters for measuring position on a circle of radius R are the angular displacement $\Delta\theta$, usually given in radians, and arc length, $s_{\text{arc}} = R\,|\Delta\theta|$.

Motion at constant speed on a circular path is called uniform circular motion. An important concept here is the associated acceleration, called centripetal acceleration, which is always directed towards the centre of the circle. This result was obtained by applying the definition of acceleration as the derivative of the velocity, and is quite counter-intuitive; you could hardly have guessed it. This illustrates the enormous power of mathematics and the importance of making precise definitions.

Another type of periodic motion described in Chapter 3 is simple harmonic motion (s.h.m.), a to-and-fro motion along a straight line. If a particle moves in uniform circular motion in the xy-plane then the x- and y-components of its position vector both vary with s.h.m. with the same amplitude and frequency but with a phase difference of $\pi/2$. Conversely, uniform circular motion in the xy-plane is equivalent to the combined effect of the two simple harmonic motions along the x- and y-axes.

This is why similar kinds of parameters are used to characterize both kinds of motion. For example, both circular motion and s.h.m. are characterized by a period T, or a parameter $\omega = 2\pi/T$ called the angular speed of uniform circular motion, or the angular frequency of s.h.m. Other important parameters of s.h.m. are the amplitude A giving the range of the motion on either side of the midpoint, and the phase constant, or initial phase, ϕ, which determines where the particle is at time $t = 0$.

Simple harmonic motion is described mathematically using the trigonometric functions, sine and cosine, which are generalizations of the trigonometric ratios used in Chapter 2 for resolving vectors. The two trigonometric functions, $\sin\theta$ and $\cos\theta$, are the x- and y-components of the position vector of a particle on a unit circle, θ being the angular coordinate measured anticlockwise from the x-axis.

Both $\sin\theta$ and $\cos\theta$, are periodic functions with period 2π. They are related to one another simply by a phase difference of $\pi/2$; if you shift the graph of $\cos\theta$ to the right by $\pi/2$ you get the graph of $\sin\theta$. Trigonometric functions have applications in almost every branch of physics and you will need to use them throughout this course.

It would be difficult to overestimate the importance of simple harmonic motion in physics. Firstly, it is possible to consider *any* periodic motion as a combination of two or more simple harmonic motions of suitable amplitudes, phases and frequencies — you have seen the simple example of uniform circular motion being equivalent to two simple harmonic motions at right angles. You will come across other examples of this idea later in the course. Secondly, the mathematical description of s.h.m. in terms of sine or cosine functions occurs in a variety of guises throughout physics, especially in the descriptions of alternating electric currents and all kinds of wave motion. In fact you will see that the concepts of angular frequency, amplitude and phase are central to our understanding of optical interference effects and much of quantum physics. This is why we spend so much time on s.h.m. in this book.

Finally, Chapter 3 describes the periodic orbital motion of bodies such as satellites, spacecraft and planets that move under the influence of point-like sources of gravitational attraction, such as the Earth or the Sun. The orbits of these bodies are ellipses. Very often the ellipses have very small eccentricity e, i.e. they are very nearly circular. The main features of orbital motion are summed up in Kepler's laws, and a topical application is the use of Hohmann transfer orbits in interplanetary travel.

Elliptical orbits can also occur in other contexts. Early theories of the structure of the hydrogen atom assumed that the negatively charged electron moves by electrical attraction in an orbit about a positively charged nucleus. This orbital motion is also described by Kepler's laws, since both electrical attraction and gravitational attraction vary inversely with the square of the distance. In a later book, *Quantum physics: an introduction*, you will see the first attempt at such theories, the Bohr model of the hydrogen atom where the orbits are assumed for simplicity to be circles.

Throughout this book you have seen the importance of making approximations and assumptions when describing phenomena or solving problems. Phenomena in the real world are often incredibly complex and impossible to describe exactly, and so the first step in solving any problem is to simplify it by making appropriate assumptions and approximations, i.e. by making a simplified model. The objective is to simplify as much as possible without throwing away the essential features of the problem. In Chapters 1 to 3 we have modelled complex objects like cars, athletes, trains, planets, etc., as point-like bodies called particles. The shape and size of the object, its composition and internal structure, etc., have all been ignored; we've shrunk the object down to a geometrical point — a drastic simplification. Yet we've been able to make important predictions such as how the range of a projectile depends on the initial speed and direction, and what the angle of projection should be for maximum range. Of course, the accuracy and usefulness of our predictions are limited by the assumptions and approximations of the model. Thus the particle model can't be used to describe many important aspects of motion such as the flexing of an athlete's body, the rotation of a planet about its axis, the spin and elastic vibrations of a struck golf ball, or the effects of air resistance. If we want to know more about these aspects of the motion then we would have to replace the particle model by one that includes the properties of the real bodies that are relevant to the problem at hand. You will see an example of this in the next book where we introduce a *rigid body*, a model for describing, for example, spinning tops, flywheels, levers, etc., where the size and shape of the bodies are essential features of the problem.

Knowing what assumptions and approximations to make and what models of the real world are appropriate in any problem is a skill you will acquire gradually as the course progresses. Be aware of this process; it's at the very heart of doing physics.

3 Introducing *Physica*

3.1 Advice about *Physica*

Physica is a powerful computing package which can be used throughout your study of *The Physical World*. It provides help with many aspects of the course, including problem-solving.

From the outset, it is important to realize that *Physica* is more open-ended than the other multimedia packages associated with the course, and should therefore be approached in a somewhat different way. Unlike a book, which is read cover-to-cover, there is no beginning or end to *Physica*, and no prescribed route through it. Instead, you can think of *Physica* as a collection of resources and tools, — a sort of super-calculator and information store — which can be used for your own purposes, as and when you need it.

Before you have the confidence to use *Physica* in this flexible way you will need to spend time learning how to use it. Because *Physica* is extensive and powerful, several hours will be needed, but this time will not be wasted. Although it is challenging to learn a new computer package, the ability to do so is becoming an increasingly important skill throughout science. So learning how to use *Physica* is a valuable learning experience in itself. More importantly, once you have become familiar with the package, you will have access to a set of valuable resources which will help with the course, and which can be used more widely.

A major aim of this chapter is to introduce *Physica*, show you some of the things it can do, and give you practice at using it for a variety of purposes. Before getting immersed in these details, we begin with a brief overview. There are four main ways of using *Physica*, illustrated by the pyramid in Figure 4.1.

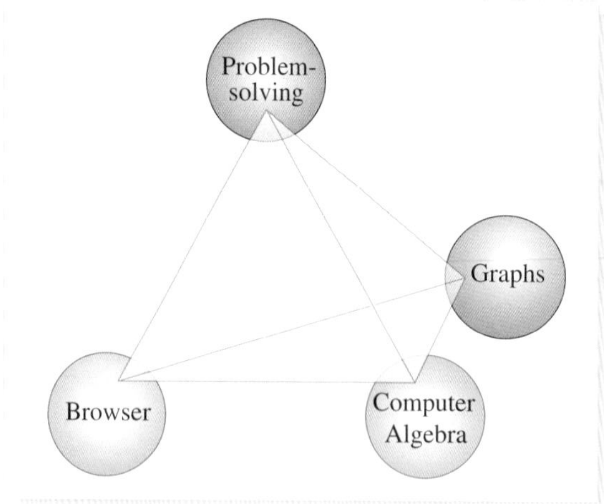

Figure 4.1 The four main ways of using *Physica*.

The *Browser* is a large data bank which you can use to look up information about physics and related mathematics. The Browser uses *hyperlinking* (don't worry if you don't understand this term yet) to connect one term to another, so you can easily follow chains of related ideas. Having followed such a chain, you can always retrace your steps, returning to your point of departure. Items in the Browser are stored under different headings, rather like documents in different drawers of a filing cabinet. The main headings, 'Physics' and 'Maths', provide an extensive glossary of terms in physics and related mathematics. Other headings provide on-line reminders on how to use *Physica*, and information directly related to the books in the course.

At the heart of *Physica* lies a set of *computer algebra tools*. Computer algebra starts from the idea that any algebra done accurately with pencil and paper must follow precise rules. These rules can be incorporated into computer programs so that computers can carry out the algebra automatically. Similar ideas apply to other mathematical operations, including differentiation.

The labours of Charles Delauney illustrate the need for automation. In 1860, Delauney was greatly relieved to publish his *Theory of the movement of the Moon*. Chapter 4 of this book contains a formula filling 128 pages, the result of twenty years work (ten years obtaining the formula, and ten more years checking it). Modern physics contains calculations even longer than Delauney's. The results of these calculations can often be stated quite briefly but their intermediate working is of mind-numbing complexity, far more suitable for computers than human brains. As part of your exposure to physics and its methods, we therefore want you to gain some experience of computer algebra.

Several different computer algebra programs have become commercially available in the last decade or so. *Physica* is based on one of the most famous, *Mathematica®*, produced by Wolfram Research. *Mathematica* is a large and powerful program incorporating well over a million lines of code. Unfortunately, *Mathematica* is just too large and intimidating for our needs, requiring knowledge about computing that would take us too far from the central focus of this course. That is why we have developed *Physica* as a major adaptation of *Mathematica*. Our aim has been to provide a series of tools that are relatively easy and straightforward to use in basic physics calculations.

The first problem we face, in using a computer algebra program, is to enter information, such as the equations we want to manipulate. It is always possible to do this via the keyboard, but this is time consuming and inevitably requires special syntax, in order to cope with notation such as derivatives, subscripts or Greek letters.

Physica provides ways of easing this burden. For example, important physics equations are arranged into menus and can be easily accessed by clicking the mouse. The Browser also collects the main equations in each book chapter, and again allows them to be accessed directly, using the mouse. So, if you want to use *Physica* to investigate or check any of the algebraic manipulations you see in the books, you should be able to obtain the relevant equations quite easily in *Physica*.

Once entered, equations can always be manipulated. For example, an equation can be rearranged so that a chosen variable is isolated on the left-hand side, or a common variable can be eliminated from two or more equations. Alternatively, both sides of an equation can be differentiated. You can also carry out manipulations of vector equations. As you will see later, two- and three-dimensional graphs can be plotted to visualize the meaning of equations. At the most basic level, you can just enter values for variables and evaluate equations, much as you would with a calculator, but with the inclusion of units for physical quantities.

Most of the operations work as single actions, giving you the final result, but not showing the intermediate working. As a matter of fact, if we revealed what was happening inside the computer, you would probably not want to know — it is usually very lengthy and quite distant from conscious human thought. For many purposes it is good enough to know that the computer gets the right answer. However, if you really want to explore the steps needed to rearrange an equation, a special tool allows you to proceed step-by-step. We recommend using this tool, in conjunction with specific equations from the course, if you have trouble following the algebra between the lines of the course texts.

Before going any further, some warnings and advice about the use of computer algebra within *Physica* are appropriate.

- Do not expect *everything* to be easier, or quicker, using *Physica* rather than pencil and paper. In order for the computer to help, you will have to give it precise information, some of which might be left implicit when working with pencil and paper. The effort needed to input the information is compensated by the ease with which the computer carries out the routine algebra, securely and without error. So *Physica* will be quicker if the time gained on the algebra is greater than the time spent on inputting information.
- One concern that is sometimes expressed about computers and calculators is that they lead to de-skilling. This is not really fair because the skills needed to use a computer are becoming increasingly important, and should certainly not be dismissed. Nevertheless, people are sometimes spotted using calculators to add 9 to 7, prompting questions about the undermining of basic skills. Similar questions arise with computer algebra. Most physicists would agree that the ability to do simple algebra, *without* the aid of a computer, is essential. In fact, for many purposes, computer algebra is just too cumbersome, and too restrictive for the task in hand. Our advice, then, is to use *Physica* in a way that supports, rather than avoids, the development of algebraic skills. Depending on your background, you may want to use the step-by-step tool mentioned above. At the very least, you should check, from time to time, that you could supply the algebra that is being done by *Physica*. Remember that you will not be able to take a computer into the exam, and will be at a serious disadvantage if you are uncertain or slow at performing basic algebraic operations with pencil and paper.
- Although *Physica* gives the right answers, its 'taste' is sometimes odd; after some manipulations, you may find that the ordering or grouping of variables in equations is different to your expectations. For example, you might expect an equation of the form $s = ut$ and get instead the equivalent equation $s = tu$. This is unavoidable,

because, once *Physica* gets seriously involved in solving and rearranging equations, it cannot understand or interpret the answers in the way that you do. As usual, computers are amazingly clever and remarkably stupid, at the same time. In practice, this should not cause any difficulty. At any rate, you should not think that *Physica* knows best in this respect. Always be prepared to rearrange its answers to suit your own taste before writing them down on paper.

3.2 Features of *Physica*

The *Graph Plotting Tools* will help you visualize the meaning of equations. Given an equation linking one variable to others, you can choose to plot a variety of graphs. For example, the equation

$$s_x = u_x t + \tfrac{1}{2} a_x t^2 ,$$

can be used to plot a graph of s_x against t. *Physica* will make certain assumptions before plotting this graph. For example, it will assume some fixed values for u_x and a_x, and a range of values for t, based on any information you have supplied. You will get the chance to alter these assumptions, and so modify the plot. You can also choose additional options, such as indicating a specific point on the graph, drawing a tangent to the graph, or shading the area under the graph. It is also possible to plot three-dimensional graphs, such as that showing how s_x depends on u_x *and* t. Similar information can also be represented on a contour plot.

There are also basic tools for visualizing numerical data, such as would be gathered from an experiment. These tools allow you to enter simple data and fit it with a straight line or a curve of a given type.

Problem-solving One of the main functions of *Physica* is to provide a wide collection of problems for you to tackle using the tools of computer algebra. The problems are classified according to the books in the course, and are further divided into topics, such as projectile motion or uniform circular motion. In the *Consolidation and skills development* chapters that end each book, you will be asked to tackle a selection of these problems. We certainly don't expect you to tackle every problem, and it will be up to you to make a choice, based on your own needs. Since the problems span the main topics of the course, it is sensible to try one or two from each category, and then decide where more practice is needed.

Various levels of help and guidance accompany the problems. The least interactive, and most supported level is called the *Guided Solution mode*. Here, you can simply watch problems being solved, step by step. Of course, to learn from this process, you will have to pause at each step and try to see why that step is being taken. If you do this, *Physica* will provide a useful source of worked examples, from which you can learn, by observation.

However, we want you to concentrate on a more active approach than this, by solving problems for yourself. You can build up to this active mode of working from within the Guided Solution mode because, at any moment, you can choose to take over and carry out the recommended step for yourself. Eventually, when you are sufficiently confident, you will be able to tackle problems completely by yourself. You can then choose whatever actions you want, but will always be able to get help from a menu, if you get stuck.

It is important to understand *why* we are asking you to solve problems using *Physica*. You might hope that the computer would make life easier, and allow you to rattle off many problems in a short time. In your early stages of learning how the interface works, this will not be true. It will take longer to solve a simple problem using the computer than with pencil and paper. Things will speed up later on, but our main aim in asking you to solve problems with *Physica* has little to do with savings (or not) in time. The experience of solving a problem on the computer is distinctly different from that obtained with pencil and paper, and is uniquely valuable.

This returns to a point mentioned earlier. Computers have to be told explicitly what to do. Having read the problem wording, you will need to collect together an appropriate set of equations. You will also need to specify any relevant known values, and choose variables for the things you are asked to find. The collection of information of this type is known as the Preparation stage in problem-solving, and is something we strongly recommend throughout the course. Because *Physica* includes the Preparation stage as an essential part of telling the computer how to proceed, it will help you develop good habits, and become well organized at the outset of tackling problems. Once sufficient information has been assembled, the solution moves into the Working stage, where the algebra is done. Here of course, *Physica* offers great advantages of speed and safety. As the details will be handled automatically, you will be able to think about algebra from a more strategic perspective. For example, you will be able to concentrate on whether there are enough equations to determine the unknown variables you want to find. Finally, it is always a good idea to check your answer to see that it makes sense. *Physica* provides some tools that can help you do this, by examining the trends of the answer as different variables change, or by looking at limiting cases.

One other aspect of *Physica* is worth mentioning, though is not recommended for everyone. The *Text Tool* allows you to produce word processed documents. The word processing features are limited by the standards of modern word processing packages, but there is one feature which is especially useful to physicists: mathematical expressions can be input with relative ease. Some students like to word process their answers to assignments but, depending on the package used, may find it difficult to input equations, especially those including subscripts, fractions, vectors, derivatives and Greek letters. The text tool can be used to produce a more professional end-result. Nevertheless, we emphasize that word processing is entirely optional and will gain no extra credit in this course. For many people, it will be an unhelpful diversion. If you find it quicker and more convenient to handwrite your answers (legibly), please continue to do so. And, if word processing takes any extra time, please consider carefully whether that time would be better spent reading the books, doing more exercises, or exploring other aspects of *Physica*.

3.3 Getting started on *Physica*

When getting to grips with any new computer package, there is a temptation to click buttons or pull down menu items immediately, before reading, and digesting, the instructions. When pressed for time, it is certainly tempting to do this, relying on the intuitive nature of the interface to do the right thing. Needless to say, this often leads to frustration. Many well-known computer packages are used ineffectively because their most powerful features are left undiscovered by such tactics. Our first advice, therefore, is to read the manual — in this case, the supplementary booklet *Using Your Computer*. If you are new to computing, this booklet contains basic advice on using a computer, which can be skipped if you are a more experienced user. But whatever your experience, please read the section: *Running Physica*.

4 Review of skills

As you work your way through this course you will acquire and develop a variety of skills. Some of these skills are specific to particular tasks, such as the ability to measure the gradients of graphs or to solve quadratic equations. Other skills needed in physics, such as seeing your way through a problem, making suitable assumptions and approximations and using appropriate physical laws, are more broadly based and involve the integration of many specific skills and lots of experience at working problems. You will acquire these broadly based *physics skills* gradually as you work your way through the course and build up a repertoire of specific skills and experience at using them. In this section we review some of the specific skills introduced in this book. We can classify them broadly into: *mathematical skills* and *information technology (IT) skills*.

4.1 Mathematical skills

After studying this book you should have skills in the following areas of mathematics and be able to carry out the specific bulleted tasks.

Gradients and derivatives

- Present data in graphical form and measure the gradient of a graph at any point.
- Use linear functions, quadratic functions and trigonometric functions to describe relationships between variables.
- Use derivatives of the above functions to describe gradients and rates of change.

Here we are building on the graphical skills you acquired from your earlier courses or other work, with an emphasis on the *gradient* of the graph and its mathematical representation, the *derivative* of the function. You should know how derivatives are obtained by using the standard derivatives in Table 1.6.

Vectors

- Recognize vector quantities and use vector notation.
- Add two vector quantities of the same kind using the triangle rule, or equivalently, by adding the corresponding components.
- Multiply a vector by a number (or any scalar).
- Resolve or project a given vector into perpendicular components; conversely, determine the magnitude and direction of a vector knowing its components.

This book may be your first introduction to vectors. You must be able to distinguish between scalar and vector quantities and discipline yourself to underline vector symbols (or to type them in bold). You will be using vectors and learning more about them throughout this course.

Trigonometry and trigonometric functions

- Convert between radians and degrees.
- Relate the sides of a right-angled triangle to one another using the trigonometric ratios $\sin\theta$ and $\cos\theta$.
- Sketch the graphs of the trigonometric functions $\sin\theta$ and $\cos\theta$, where θ is any number.
- Sketch the graphs of the sinusoidally varying functions $A\sin(\omega t + \phi)$ and $A\cos(\omega t + \phi)$, where A, ω and ϕ are constants.
- Determine the derivatives of sinusoidally varying functions.

You have probably met the trigonometric ratios in your previous studies. You must now become familiar with their generalizations, the trigonometric functions, since they will be used throughout this course. Make sure you know how the graphs of sinusoidally varying functions depend on the values of the constants A, ω and ϕ.

Most physicists find it useful to memorize the sines, cosines and radian measures of at least the common angles, 0°, 30°, 45° and 90°; it is also useful to memorize the derivatives of the sine and cosine functions.

Algebra

- Read and manipulate algebraic expressions and equations.
- Solve quadratic equations and select the appropriate solution.

Algebra is used to give precise and succinct statements of relationships between variables. Reading an equation means being able to say in words what the relationship is, or conversely, being able to express a given relationship succinctly as an algebraic expression.

We manipulate algebra and equations whenever we want to express the same relationships in a different way, such as changing the subject of an equation. We do this by rigorously applying the rules of algebra, which are all variations on the theme: 'what you do to one side of an equals sign you must also do to the other side'. Do not expect algebraic skills to come quickly; you will develop them slowly as you progress through the course and gain experience at using them.

Quadratic equations occur throughout physics and are common when we ask a question that has two possible answers. This happens surprisingly often, so you must know how to solve quadratic equations using the quadratic equation formula or in simple cases by factorization. Sometimes the physical interpretation of the two solutions is obvious. Sometimes it isn't, and you have to think carefully about which solution is the answer you are looking for.

4.2 IT skills

After studying this book (including the various multimedia activities that it incorporates) you should have a number of IT skills and be able to carry out the bulleted tasks

- Start and navigate various multimedia packages.
- Input information using mouse and keyboard in response to appropriate prompts.
- Perform simple mathematical operations and evaluations using the *Physica* package.

5 Basic skills and knowledge test

You should be able to answer these questions without referring to earlier chapters. Leave your answers in terms of π, $\sqrt{2}$, etc. where appropriate.

Question 4.1 Refer to Figure 4.2. What is the position of point P and the displacement of point P from point Q?

Question 4.2 What is the velocity of a particle that takes 5 s to move with constant speed from point Q to point P in Figure 4.2?

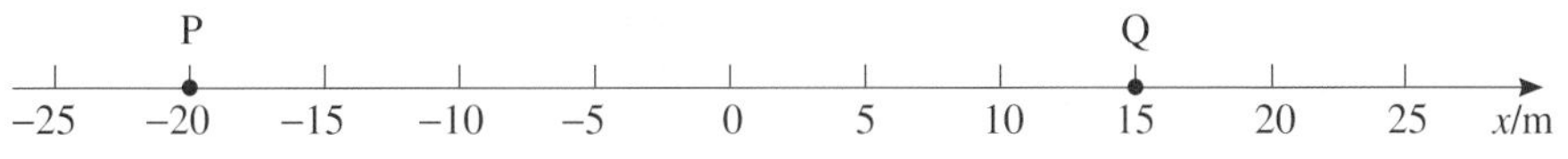

Figure 4.2 For use with Questions 4.1 and 4.2.

Question 4.3 Let $\boldsymbol{a}$ denote a displacement vector of magnitude 5 m pointing vertically upwards. State the magnitudes and directions of the following three vectors: $2\boldsymbol{a}$, $-\boldsymbol{a}$, $-0.1\boldsymbol{a}$.

Question 4.4 Vectors $\boldsymbol{b}$ and $\boldsymbol{c}$ both have magnitude 3 m and lie in a horizontal plane. Vector $\boldsymbol{b}$ points due east and vector $\boldsymbol{c}$ points due north. State the directions of the two vectors $\boldsymbol{b} + \boldsymbol{c}$ and $\boldsymbol{b} - \boldsymbol{c}$.

Question 4.5 Determine the x- and y-components of the following vectors which all lie in the xy-plane: (a) vector $\boldsymbol{a}$ in Figure 4.3a; (b) vector $\boldsymbol{b}$ of magnitude 3 m in Figure 4.3b; (c) vector $\boldsymbol{c} = (3, -6)$ m (not shown in any figure.)

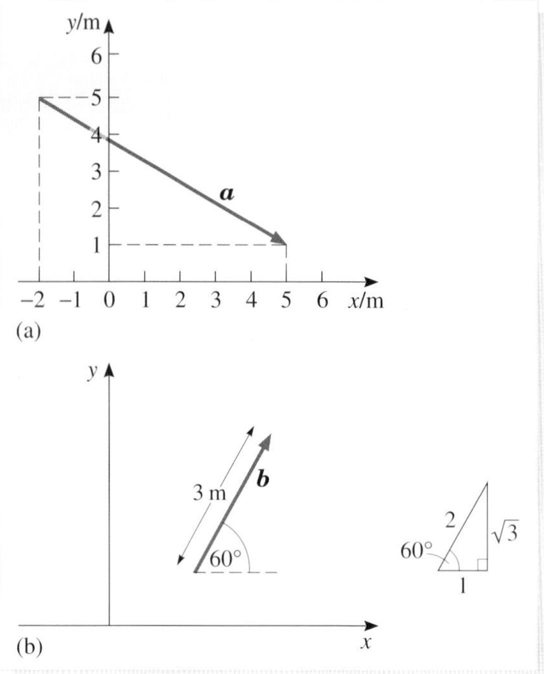

Figure 4.3 For use with Question 4.5.

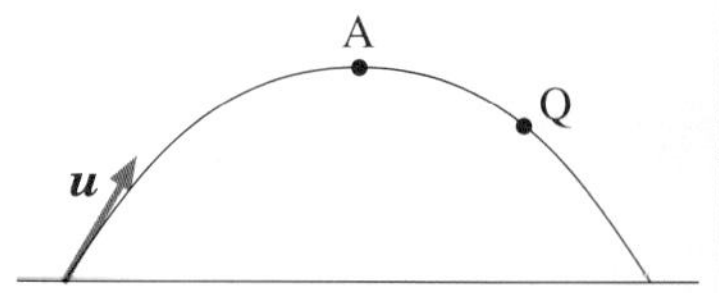

Figure 4.4 For use with Question 4.7.

Question 4.6 Determine the magnitudes of (a) the vector $\boldsymbol{c} = (3, -6)$ m, and (b) the vector $\boldsymbol{u} = \sqrt{2}\,(1,1)\,\mathrm{m}$.

Question 4.7 Figure 4.4 shows the parabolic trajectory of a golf ball. What is the acceleration of the ball at any point Q on the trajectory?

Figure 4.5 For use with Question 4.8.

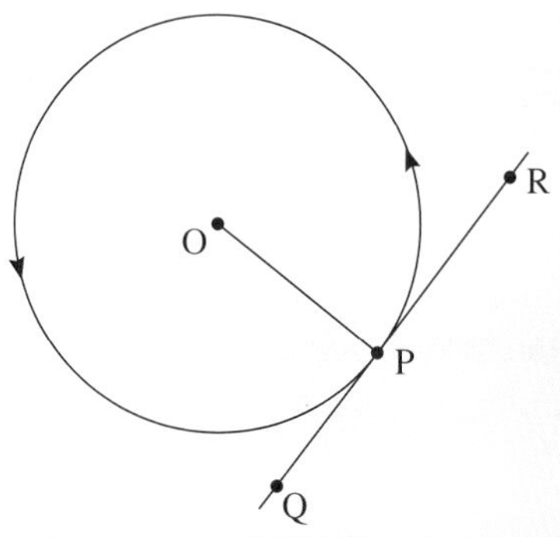

Question 4.8 The particle at point P in Figure 4.5 moves anticlockwise at constant speed around the circle. State, in terms of points and lines shown on the figure, the direction of the velocity of the particle, and the direction of the acceleration.

Question 4.9 A particle moves on the x-axis in simple harmonic motion between end points $x = -2$ m and $x = 2$ m. (a) What is the amplitude of the motion? (b) What is the period if it takes 1 s to move from the origin to $x = 2$ m?

Question 4.10 Refer to the particle in Question 4.9. (a) At what point is the particle's speed largest? (b) At what point is the acceleration zero?

Question 4.11 The displacement of a particle moving in simple harmonic motion is given by $x = A\sin(\omega t + \phi)$. The displacement–time graph for the particle is shown in Figure 4.6. State the angular frequency and the initial phase. (The graph of $\sin\theta$ is also shown for reference.)

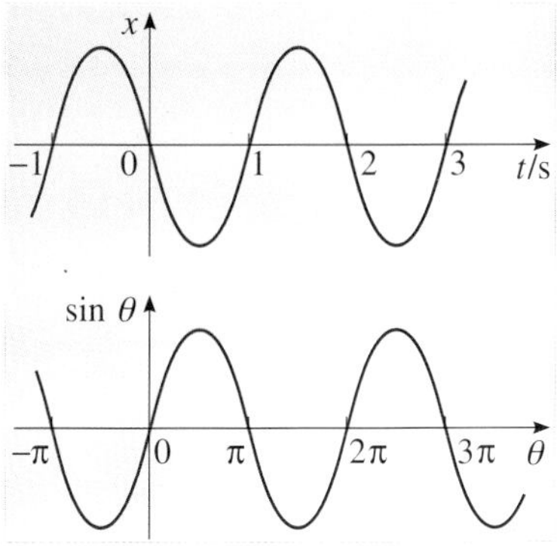

Figure 4.6 For use with Question 4.11.

Question 4.12 Let $\boldsymbol{r}$ be the position vector of a particle moving in space. Define (a) the velocity of the particle and (b) the acceleration of the particle, in terms of $\boldsymbol{r}$.

Question 4.13 At what point on its orbit, relative to the Sun, does a planet move fastest?

Question 4.14 How many radians is 40°? How many degrees is π radians?

Question 4.15 A particle moves through an angular displacement of $\pi/2$ radians on a circle of radius 5 m. What arc length has it moved through?

Question 4.16 The curve in Figure 4.7 is the position–time graph for a particle moving along the x-axis. Estimate the gradient of the graph at $t = 23$ s and state its physical significance.

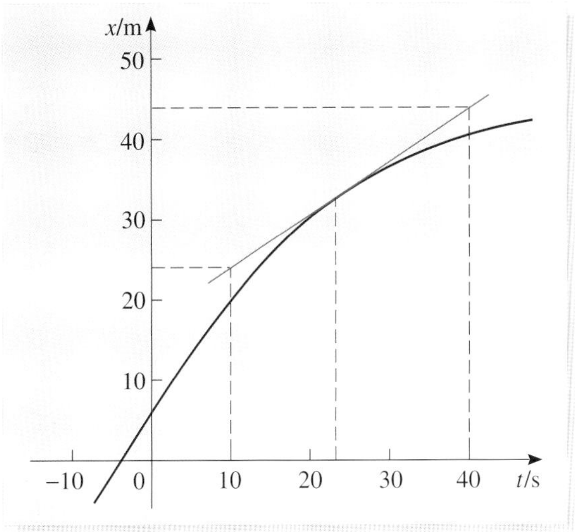

Figure 4.7 For use with Questions 4.16 and 4.17.

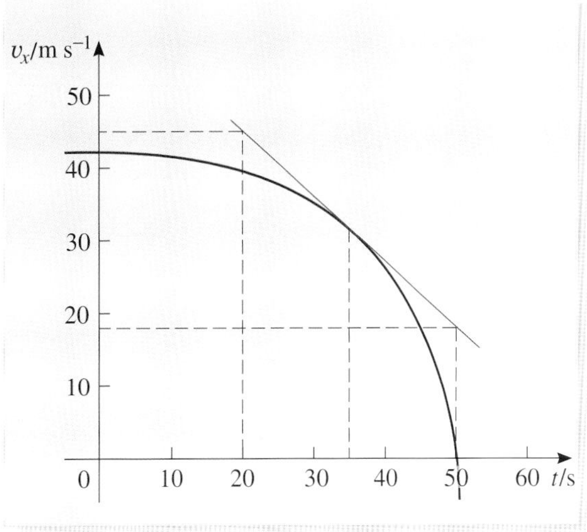

Figure 4.8 For use with Questions 4.18 and 4.19.

Question 4.17 Refer again to Figure 4.7. Where is the particle at time $t = 0$ s? At what time does the particle pass through the origin?

Question 4.18 The curve in Figure 4.8 is the velocity–time graph of a particle moving along the x-axis (not the same particle as in Question 4.16). Estimate the gradient of the graph at $t = 35$ s and state its physical significance.

Question 4.19 Refer again to Figure 4.8. How fast is the particle moving at time $t = 0$ s? At what time is the particle instantaneously at rest?

Question 4.20 What are the solutions of the following three quadratic equations: (a) $(x-3)(x+2) = 0$; (b) $(x-1)^2 = 0$; (c) $x(x+5) = 0$?

(You should not need to use the quadratic equation formula to answer this question; just ask yourself what values of x make the left-hand side of the equation zero.)

Question 4.21 Three scalar variables a, b and c are related by the expression $a = \dfrac{5\sqrt{b}}{c^2}$. How does a change (i) when b doubles in value while c stays fixed, (ii) when c triples in value while b remains fixed? (iii) If b increases by a factor of 4, by how much must c change if a is to remain unchanged?

Question 4.22 The displacement of a particle moving on the x-axis is given by the quadratic function $x(t) = At^2 - Bt + C$, where A, B and C are constants and t is time. The derivatives of x are

$$\frac{dx}{dt} = 2At - B \quad \text{and} \quad \frac{d^2x}{dt^2} = 2A.$$

Where is the particle at time $t = 0$, and what are its velocity and acceleration at $t = 0$? ■

6 Interactive questions

Open University students should leave the text at this point and use the interactive questions package for this book. When you have completed these questions, you should return to the text. You should not spend more than 3 hours on this package.

7 *Physica* problems

Open University students should leave the text at this point and tackle the *Physica* problems that relate to this book. When you have completed these questions, you should return to the text. You should not spend more than 4 hours on this package.

Answers and comments

Q1.1 Your list might well include items such as: the motion of a passenger on a train, or in a plane or in any other vehicle, as long as it is the passenger's overall position that is important, and not their posture or internal movement. You might also have listed the vehicles themselves, provided the same conditions apply. Indeed, you might list almost anything, including the Earth or the Sun, provided you are considering a context in which the moving object can be treated as point-like.

Q1.2 (a) From Figure 1.6, the position at $t = 32$ s is 59 m. (b) In Figure 1.5 the car is at $x = 30$ m. According to Figure 1.6, the time corresponding to $x = 30$ m is 21 s.

Q1.3 The whole graph could be shifted to the right or the left (without altering its shape) by choosing the origin of time ($t = 0$ s) to be earlier or later than that used in Table 1.2.

Q1.4 Since we are interested in the displacement of the getaway car from the police car, we need to plot $x_{get} - x_{pol}$. You may find it useful to tabulate these values by adding another column (headed $x_{get} - x_{pol}$) to Table 1.3. The result of the plot is given in Figure 1.37.

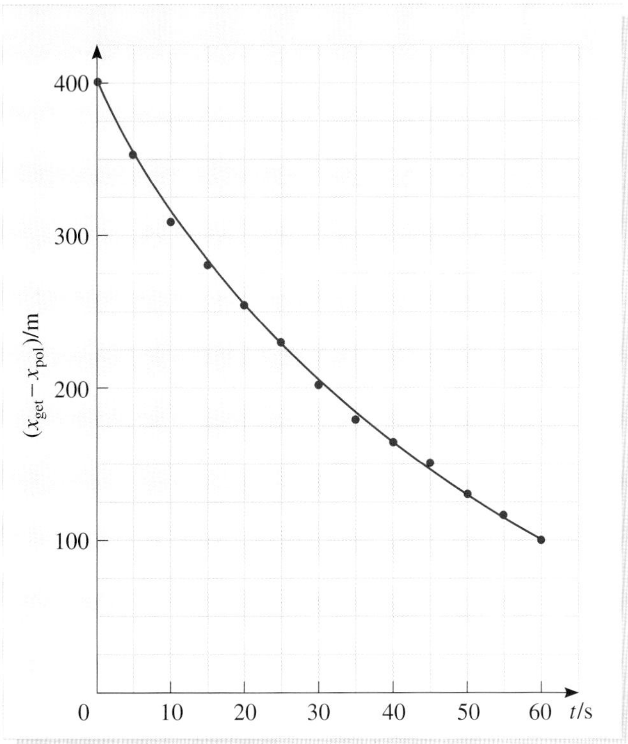

Figure 1.37 A displacement–time graph based on Table 1.3.

Q1.5 Only you will know the answer to this, but it is common to see graphs in which the axes have not been labelled, or the units have been omitted. This is especially true when automated graph-plotting packages are used; such packages often require special instructions if they are to show labels and units, and these are easily overlooked. If you are using such a package (or a graphical calculator), don't forget that the line you have to plot is far from being the whole graph: axes and labels are also important.

Q1.6 The displacement of the car from the pedestrian is plotted in Figure 1.38. The significant feature is that we get a straight line and consequently the relative motion is uniform, i.e. at constant speed. The straight line does *not* run through the origin, so this curve is of the general form $x = At + B$, where x stands for $x_{car} - x_{ped}$ in this case.

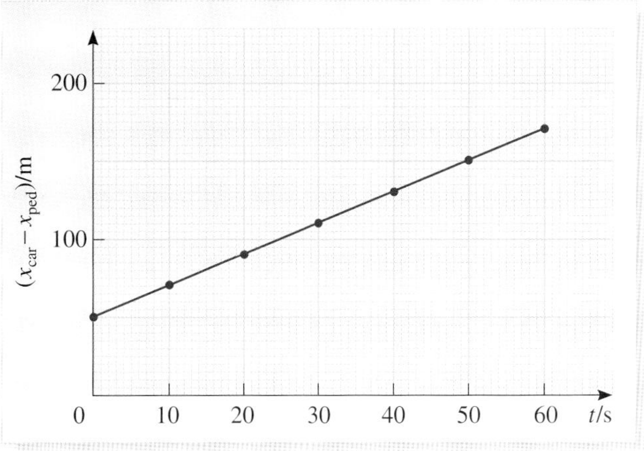

Figure 1.38 The result of plotting the displacements in Table 1.5.

Q1.7 Using $v_x = \dfrac{x_2 - x_1}{t_2 - t_1}$ and Figure 1.10b we obtain

$$v_x = \frac{(210 - 180)\,\text{m}}{(60 - 50)\,\text{s}} = 3\,\text{m}\,\text{s}^{-1}.$$

A point to note here is that choosing such closely separated values as $t = 50$ s and $t = 60$ s makes the evaluation of the velocity trickier, and more prone to error than would have been the case if more widely separated values had been chosen. When evaluating agradient from a graph, it is always wise to use the widest convenient range of values on the horizontal axis.

Q1.8 (a) The speed is given by the magnitude of $\Delta x/\Delta t$ obtained from the velocity–time graph. In the order of increasing speed, the objects are B, D, A, C. (b) The objects with positive velocity have positive values of $\Delta x/\Delta t$, these are A and B. (c) Treating large negative values as being *less* than small negative values, and any negative value as less than any positive value, as is conventional, the list is C, D, B, A.

Q1.9 The gradient of the temperature–height graph is

$$\frac{\Delta T}{\Delta h} = \frac{T_2 - T_1}{h_2 - h_1}.$$

With $h_1 = 0$ and $h_2 = 2\,\text{km}$, Figure 1.12 shows that $T_1 = 20\,°\text{C}$ and $T_2 = 6\,°\text{C}$.

It follows that the gradient is

$$\frac{\Delta T}{\Delta h} = \frac{T_2 - T_1}{h_2 - h_1} = \frac{(6-20)\,°\text{C}}{(2-0)\,\text{km}} = -7\,°\text{C km}^{-1}.$$

Note that care has to be taken to arrange the values in the correct order when performing the subtractions if the right sign (minus in this case) is to be obtained.

Q1.10 The equation will have the general form of Equation 1.6a, with velocity $\upsilon_x = 1\,\text{m s}^{-1}$ (found from the gradient) and initial position $x_0 = -20\,\text{m}$ (found from the value of x when $t = 0$). Consequently the required equation may be written:

$$x = (1\,\text{m s}^{-1})t - (20\,\text{m}).$$

Q1.11 Since υ_x = constant in this case, the motion is described by the uniform motion equation $x = \upsilon_x t + x_0$. However, on this occasion we are not given the value of x_0, though we are given enough information to work it out. Substituting $x = -2\,\text{m}$ and $t = 100\,\text{s}$ into the equation we find

$$-2\,\text{m} = (-12\,\text{m s}^{-1}) \times (100\,\text{s}) + x_0$$
$$= (-1200\,\text{m}) + x_0$$

Adding 1200 m to both sides shows that $1198\,\text{m} = x_0$.

It follows that for this particular motion the equation of uniform motion takes the form $x = (-12\,\text{m s}^{-1})t + (1198\,\text{m})$.

Consequently, when $t = 250\,\text{s}$,

$$x = -3000\,\text{m} + 1198\,\text{m} = -1802\,\text{m}.$$

An alternative method that requires slightly fewer manipulations may be based on the definition of the gradient and the given value of the velocity.

Q1.12 The relevant velocity–time graph is shown in Figure 1.39. In this case there are two rectangular areas to evaluate. Their total area is

$$(20\,\text{s}) \times (1\,\text{m s}^{-1}) + (20\,\text{s}) \times (10\,\text{m s}^{-1}) = 220\,\text{m}.$$

This is equal to the change in position over the full 40 s duration of the motion.

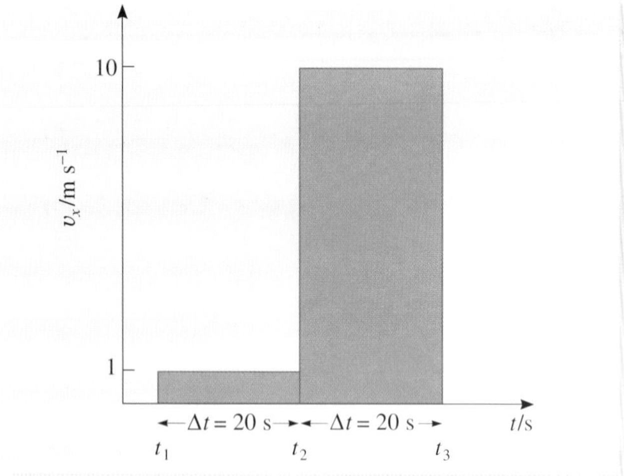

Figure 1.39 The graph for Q1.12. Note that the time values are not known, only the durations, which have been indicated by Δt.

Q1.13 The relevant tangents are shown in Figure 1.22. Their gradients give the following estimates for the instantaneous velocities

$$\upsilon_x(5\,\text{s}) = \frac{\Delta x}{\Delta t} = \frac{(8.3-0)\,\text{m}}{(15.0-2.5)\,\text{s}} = 0.67\,\text{m s}^{-1},$$

$$\upsilon_x(10\,\text{s}) = \frac{\Delta x}{\Delta t} = \frac{(13.7-0)\,\text{m}}{(15.0-5.0)\,\text{s}} = 1.37\,\text{m s}^{-1}.$$

Q1.14 (a) The particle moves in the direction of increasing x when its velocity is positive. This occurs in regions A and B. Conversely, the particle moves in the direction of decreasing x when its velocity is negative. This occurs in regions C and D.

(b) The particle is speeding up when the *magnitude* of its velocity is increasing. This occurs in regions A and C. Conversely, the particle is slowing down in regions B and D.

(c) The particle will have a positive acceleration when the gradient of the velocity–time graph is positive. This occurs in regions A and D. Conversely the acceleration is negative in regions B and C.

Q1.15 The expression for the acceleration is given by

$$a_x(t) = \frac{d\upsilon_x(t)}{dt} = \frac{d(kt^2)}{dt} = 2kt$$

using Table 1.6. Therefore $a_x(3\,\text{s}) = (2 \times 4 \times 3)\,\text{m s}^{-2} = 24\,\text{m s}^{-2}$.

Q1.16 Using Equations 1.14 and 1.15 and Table 1.6, we find

$$\frac{dx(t)}{dt} = \frac{d}{dt}(k_0 + k_1 t + k_2 t^2)$$
$$= \frac{d}{dt}(k_0) + \frac{d}{dt}(k_1 t) + \frac{d}{dt}(k_2 t^2)$$
$$= 0 + k_1 + 2k_2 t,$$

$$a_x(t) = \frac{dv_x(t)}{dt} = \frac{d}{dt}\left(\frac{dx(t)}{dt}\right)$$
$$= \frac{d}{dt}(k_1 + 2k_2 t)$$
$$= \frac{d}{dt}(k_1) + \frac{d}{dt}(2k_2 t) = 2k_2.$$

Q1.17 (a) The displacement over the first 6 s is equal to the total signed area between the graph and the t-axis and between $t = 0$ s and $t = 6$ s in Figure 1.40. Recall that regions below the axis are regarded as negative areas. In this case the total area is composed of two triangles, one above the axis, one below. The total displacement is therefore given by $s_x(6\,\text{s}) = (1/2) \times (1\,\text{s}) \times (1.2\,\text{m s}^{-1}) - (1/2) \times (5\,\text{s}) \times (6\,\text{m s}^{-1})$ i.e. $s_x(6\,\text{s}) = (0.60\,\text{m}) - (15\,\text{m}) = -14.4\,\text{m}$.

(b) The distance travelled between $t = 2$ s and $t = 6$ s will be the magnitude of the displacement over that time. Note that the displacement will be negative since it is represented by the area of the colour-shaded trapezium, which is entirely below the axis. However, the corresponding distance will be positive (since it is a magnitude) and will have the value

$s = (1/2) \times (6\,\text{s} - 2\,\text{s}) \times (1.2\,\text{m s}^{-1} + 6\,\text{m s}^{-1})$
$= 14.4\,\text{m}$.

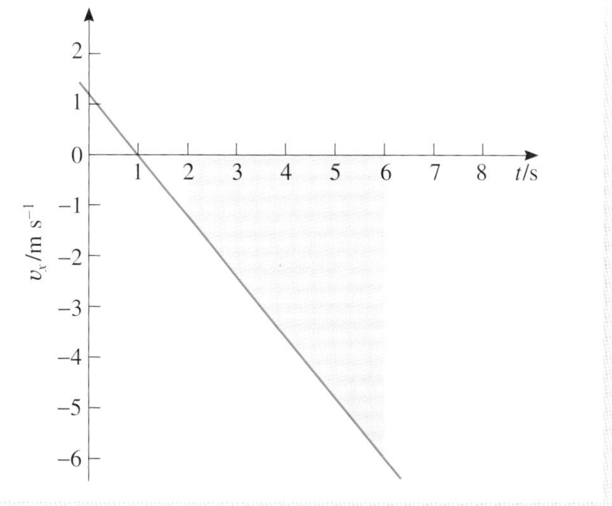

Figure 1.40 The velocity–time graph for Q1.17.

Q1.18 In this case we wish to eliminate a_x from Equations 1.28a and 1.28b. One way is to rearrange Equation 1.28b (subtracting u_x from both sides and dividing both sides by t) to obtain:

$$a_x = (v_x - u_x)/t.$$

Substituting this into Equation 1.28a gives

$$s_x = u_x t + \tfrac{1}{2}\frac{(v_x - u_x)}{t}t^2$$
$$= u_x t + \tfrac{1}{2}(v_x - u_x)t$$
$$= \tfrac{1}{2}(v_x + u_x)t.$$

Q1.19 Graphically, s_x is the signed area under the velocity–time graph between the given times. In this case those times are $t = 0$ when the initial velocity is u_x and some later time t when the velocity is v_x. Since the velocity–time graph for uniformly accelerated motion is a straight line of fixed gradient, the area required will always be either a trapezium (above or below the axis) or a pair of triangles. (Figure 1.40 shows a particular case.)

Q1.20 (a) In this case we know $s_x = 20$ m, $v_x = 30\,\text{m s}^{-1}$ and $a_x = 2\,\text{m s}^{-2}$, and we need to find u_x. As a first step we should make u_x the subject of an equation that involves the known quantities. Rearranging Equation 1.28c gives

$$u_x^2 = v_x^2 - 2a_x s_x.$$

Taking the square root of each side

$$u_x = \sqrt{v_x^2 - 2a_x s_x}.$$

Substituting the given values,

$$u_x = \sqrt{900 - 80}\,\text{m s}^{-1} = 28.6\,\text{m s}^{-1}.$$

(b) The duration of the acceleration is given by Equation 1.26 as $t = (v_x - u_x)/a_x$, which was itself obtained by rearranging constant acceleration equations. Substituting the relevant values gives

$$t = \frac{(30.0 - 28.6)\,\text{m s}^{-1}}{2.00\,\text{m s}^{-2}} = 0.70\,\text{s}.$$

Q1.21 Using Equation 1.28c

$$v_x^2 = u_x^2 + 2a_x s_x$$

with $u_x = 4.00\,\text{m s}^{-1}$, $v_x = 16.0\,\text{m s}^{-1}$ and $a_x = 2.00\,\text{m s}^{-2}$, we obtain, after rearranging

$$s_x = \frac{v_x^2 - u_x^2}{2a_x} = \left(\frac{256 - 16.0}{4}\right)\text{m} = 60.0\,\text{m}.$$

Q1.22 (a) Using a coordinate system in which up is the positive direction means that the acceleration due to gravity will be negative, that is $a_x = -g$. With $v_x = 0\,\text{m s}^{-1}$ and $s_x = 130\,\text{m}$, the initial velocity will be given (as in Q1.20) by

$$u_x = \sqrt{v_x^2 - 2a_x s_x}$$
$$= \sqrt{0-(-2\times 9.81\times 130)}\,\text{m s}^{-1}$$
$$= 50.5\,\text{m s}^{-1}.$$

(b) In this case we know $u_x = 0\,\text{m s}^{-1}$, $v_x = 50.5\,\text{m s}^{-1}$ and $s_x = 10.0\,\text{m}$. (Note that what was an initial velocity in the last part of the question has become a final velocity in this part.) Rearranging Equation 1.28c and substituting the given values

$$a_x = \frac{v_x^2 - u_x^2}{2s_x} = \left(\frac{(50.5)^2 - 0}{2\times 10.0}\right)\text{m s}^{-2}$$
$$= 128\,\text{m s}^{-2}.$$

To find the duration of the acceleration rearrange Equation 1.28b

$$t = \frac{v_x - u_x}{a}$$
$$= \frac{(50.5-0)\,\text{m s}^{-1}}{128\,\text{m s}^{-2}}$$
$$= 0.4\,\text{s}.$$

(c) The time in free motion is given by Equation 1.26 with $u_x = 50.5\,\text{m s}^{-1}$, $v_x = -50.5\,\text{m s}^{-1}$ and $a_x = -9.81\,\text{m s}^{-2}$

$$t = \frac{v_x - u_x}{a_x} = \frac{[(-50.5)-(50.5)]\,\text{m s}^{-1}}{-9.81\,\text{m s}^{-2}}$$
$$= \frac{-101}{-9.81} = 10.3\,\text{s}.$$

(Pay attention to signs here, remember that up is positive so the initial velocity will be positive, but the final velocity will be negative.)

(d) When at its highest point the acceleration of the vehicle will be $a_x = -g$. (The fact that the vehicle's velocity is momentarily zero as it passes through the highest point does not affect the (constant) acceleration.)

(e) When the particle returns to its starting point its displacement from that point will be zero. The distance travelled, however, will be twice the height of the tower, 280 m.

(f) The acceleration–time graph is given in Figure 1.41. Note that the initial and final accelerations are both positive because both increase the vehicle's velocity, even though the final acceleration reduces the vehicle's speed. (This subtlety concerning acceleration was discussed in Section 4.2.) The plan is feasible. In fact a similar system is in use at various drop-towers and drop-shafts.

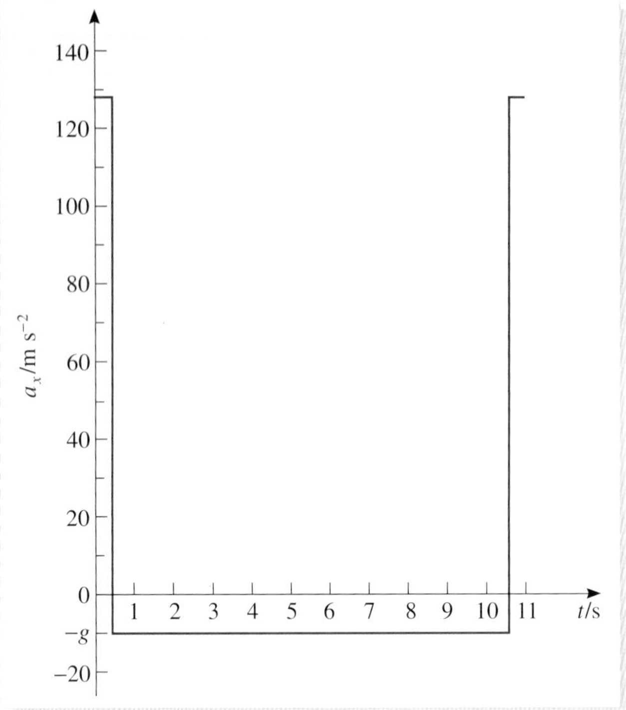

Figure 1.41 The acceleration–time graph for Q1.22(f).

Q1.23 The graph is given in Figure 1.42. Notice how the vertical axis is scaled by 10^5 in order to avoid a confusion of numbers on this axis. Since the pressure is the dependent variable, it is plotted on the vertical axis. Conversely, since height is the independent variable, it is plotted along the horizontal axis.

The rate of change of atmospheric pressure with height at 10 km is given by the gradient of the graph at this point. Using the tangent shown in Figure 1.42, we obtain

$$\frac{\Delta P}{\Delta h} = \frac{P_2 - P_1}{h_2 - h_1} = \frac{(0 - 0.565\times 10^5)\,\text{Pa}}{(16.8-0)\,\text{km}}$$
$$= -3400\,\text{Pa km}^{-1}.$$

(Differentiating the function that was used to produce Table 1.8 gives a gradient of $-3435\,\text{Pa km}^{-1}$ at 10 km.)

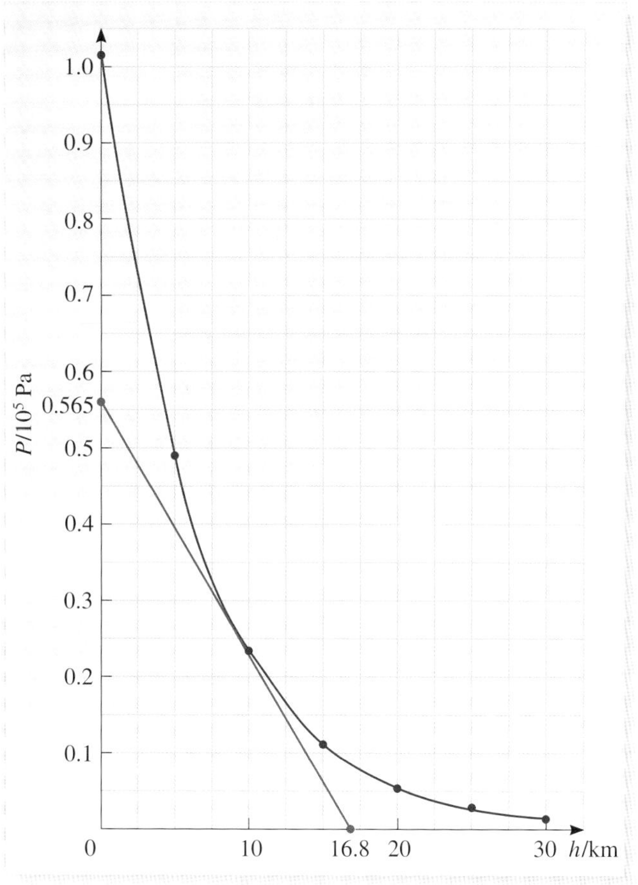

Figure 1.42 Graph for Q1.23.

Q1.24 (a) The *position* of a point on a line is represented by a coordinate x, measured from some arbitrarily chosen origin. Such a point might represent the instantaneous position of a particle moving along the line. *Displacement* refers to the difference in position of two points, these might be the initial and final positions of a moving particle, in which case the displacement would represent the change in the particle's position.

(b) The position–time graph for Table 1.9 is given by the curve in Figure 1.43.

(c) The instantaneous velocity is given by the gradient of the tangent at $t = 5$ s. The gradient of the tangent shown is

$$\frac{\Delta x}{\Delta t} = \frac{(8.5 - 0)\,\text{km}}{(10 - 1.7)\,\text{s}} = 1.0\,\text{km}\,\text{s}^{-1}.$$

(This result compares with the value of 1.087, which is obtained by differentiating the function that was used to produce Table 1.9.)

(d) The displacement–time graph for the second particle is the tangent at $t = 5$ s, which we have already drawn in Figure 1.43, since for uniform motion $x = x_0 + v_x t$. From this tangent, we can see that the position of the second particle at $t = 10$ s is 8.5 km. Hence, the displacement of the second particle from the first is (8.5 – 12.8) km = –4.3 km. Note that the minus sign is an essential part of the answer.

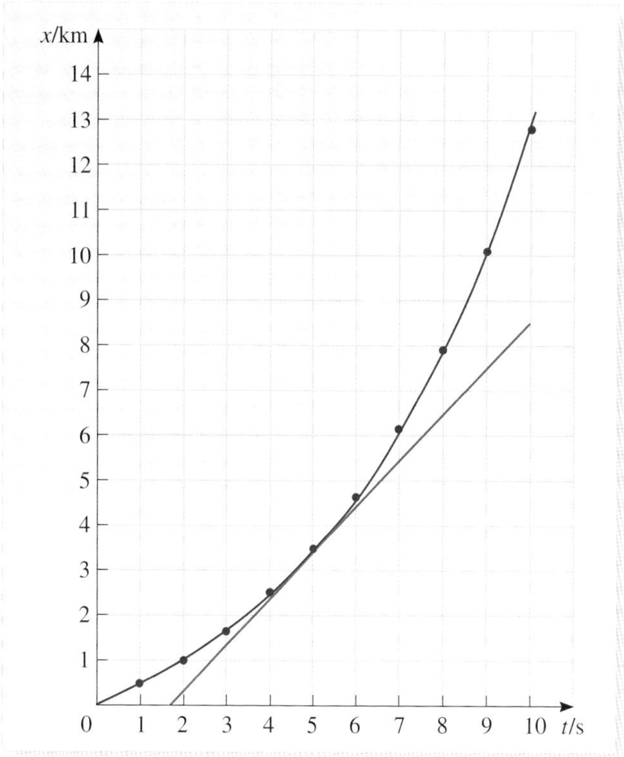

Figure 1.43 Position–time graph for Q1.24.

Q1.25 (a) The *velocity* of a particle is its rate of change of position. The *acceleration* of a particle is its rate of change of velocity.

(b) (i) The velocity–time graph for Table 1.10 is given in Figure 1.44.

(ii) The displacement is equal to the signed area under the velocity–time graph. For the interval between 0 s and 10 s, the areas above and below the time axis cancel and consequently the displacement is zero.

For the interval between 10 s and 15 s, the displacement is given by $-(1/2) \times 5 \times 4$ m = -10 m.

For the interval between 15 s and 20 s, the displacement is given by $(1/2) \times 5 \times 2$ m = 5 m.

The total displacement is the sum of all three contributions: $(0 - 10 + 5)$ m = -5 m.

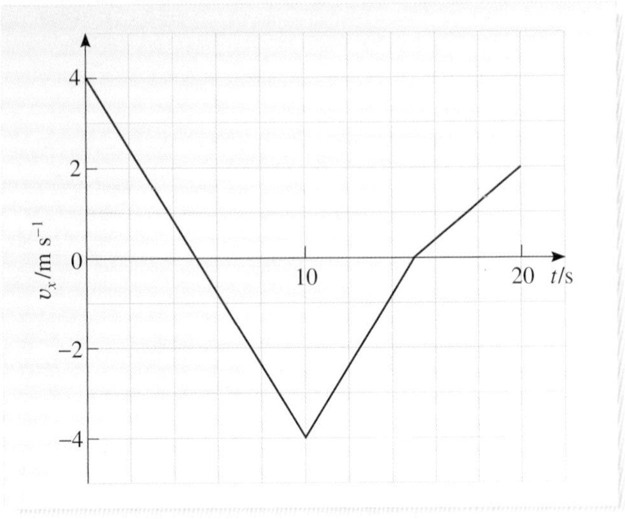

Figure 1.44 Velocity–time graph for Q1.25.

Q1.26 (a) $dz/dy = 2 + 3y^2$. (b) Substituting $y = 2$ into the derivative found in part (a) we obtain

$\frac{dz}{dy}(2) = 2 + 3 \times 2^2 = 14$. (c) Since the gradient is equal to dz/dy evaluated at $y = 2$, the answer is 14.

Q1.27 Choose an x-axis pointing vertically upwards from the Moon's surface. Using Equation 1.28b

$$v_x = u_x + a_x t$$

with $v_x = -50\,\text{m s}^{-1}$, $a_x = -1.6\,\text{m s}^{-2}$ and $t = 50$ s, we obtain

$$u_x = v_x - a_x t = (-50 + 1.6 \times 50)\,\text{m s}^{-1} = 30\,\text{m s}^{-1}.$$

Q1.28 $v_x(t) = \frac{dx}{dt} = 2At$ and therefore

$$v_x(10\,\text{s}) = (2 \times 4.0 \times 10)\,\text{m s}^{-1} = 80\,\text{m s}^{-1}.$$

Q1.29 Taking up as the positive direction, and using Equation 1.28a

$$s_x = u_x t + \tfrac{1}{2} a_x t^2$$

with $u_x = 0\,\text{m s}^{-1}$, $a_x = -9.81\,\text{m s}^{-2}$ and $s_x = -1.80$ m, we obtain

$$t = \sqrt{2s_x/a_x} = \sqrt{2 \times 1.80/9.81}\,\text{s} = 0.61\,\text{s}.$$

Q1.30 (a) Taking up as positive, and using Equation 1.28a

with $u_x = 0\,\text{m s}^{-1}$, $a_x = 2.00\,\text{m s}^{-2}$ and $t = 20.0$ s, we obtain

$$s_x = (1/2) \times 2 \times (20.0)^2\,\text{m} = 400\,\text{m}.$$

This is the displacement from the initial position when the motor fails. It follows that the height at which the motor fails is $s = |s_x| = 400$ m.

(b) Using Equation 1.28b

$$v_x = u_x + a_x t$$

with $u_x = 0\,\text{m s}^{-1}$, $a_x = 2.00\,\text{m s}^{-2}$ and $t = 20.0$ s, we obtain

$$v_x = 0 + 2 \times 20.0\,\text{m s}^{-1} = 40.0\,\text{m s}^{-1}.$$

(c) As shown in Section 5.2, eliminating t from Equations 1.28a and 1.28b gives Equation 1.28c

$$v_x^2 = u_x^2 + 2a_x s_x,$$

which we can write as

$$s_x = \frac{v_x^2 - u_x^2}{2a_x}.$$

(d) Substituting $v_x = 0\,\text{m s}^{-1}$, $u_x = 40.0\,\text{m s}^{-1}$ and $a_x = -9.81\,\text{m s}^{-2}$ in the above equation, we obtain

$$s_x = \frac{-(40.0)^2\,\text{m}^2\,\text{s}^{-2}}{-2 \times 9.81\,\text{m s}^{-2}}$$
$$= 81.5\,\text{m}.$$

Adding this result to that the displacement obtained in part (a) we obtain a total height of 482 m.

(e) Using Equation 1.28a

$$s_x = u_x t + \tfrac{1}{2} a_x t^2$$

with $u_x = 0\,\text{m s}^{-1}$, $a_x = -9.81\,\text{m s}^{-2}$ and $s_x = -482$ m, we obtain

$$t = \sqrt{2s_x/a_x} = \sqrt{2 \times 482/9.81}\,\text{s}$$
$$= 9.91\,\text{s}.$$

Q2.1 Reading from the axes, and using the units marked on the axes, the x-coordinate of the ball is 9.9 m and the y-coordinate is 1.25 m. As an ordered pair this is (9.9 m, 1.25 m).

Q2.2 The length of the arrow is 8.5 cm. The scales on both axes are such that 1 cm on the diagram represents 1 m in real space. Hence the length in real space is 8.5 m. (Notice that 8.5 cm is *not* an acceptable answer.)

Since the scaling on both axes is the same, we can measure θ with a protractor. Hence, $\boldsymbol{r}$ makes an angle of 25° with the x-axis.

Q2.3 The coordinates of A are $x = 7.7$ m and $y = 3.7$ m. The distance r is given by

$$r = \sqrt{x^2 + y^2} = \sqrt{(7.7)^2 + (3.7)^2}\,\text{m}$$
$$= 8.5\,\text{m}.$$

The direction of OA is given by

$$\cos\theta = \frac{x}{r} = \frac{7.7}{8.5} = 0.91$$

and hence $\theta = \arccos(0.91) = 25°$.

Both of these answers agree with those obtained in Q2.2, to within the accuracy with which we can make measurements on Figure 2.5.

Notice that arccos(x) means 'the angle whose cosine is x'. The appropriate function key on your calculator will probably be labelled as '$\cos^{-1}$', but it could be labelled 'arccos'.

Q2.4 The position vectors are shown in Figure 2.38. The distances from the origin are given by

$$r_A = \sqrt{1^2 + 2^2}\ \text{m} = \sqrt{5}\ \text{m}$$

$$r_B = \sqrt{1^2 + (-2)^2}\ \text{m} = \sqrt{5}\ \text{m}$$

$$r_C = \sqrt{(-1)^2 + 2^2}\ \text{m} = \sqrt{5}\ \text{m}.$$

The magnitude of the position vector is equal to the distance of the specified point from the origin.

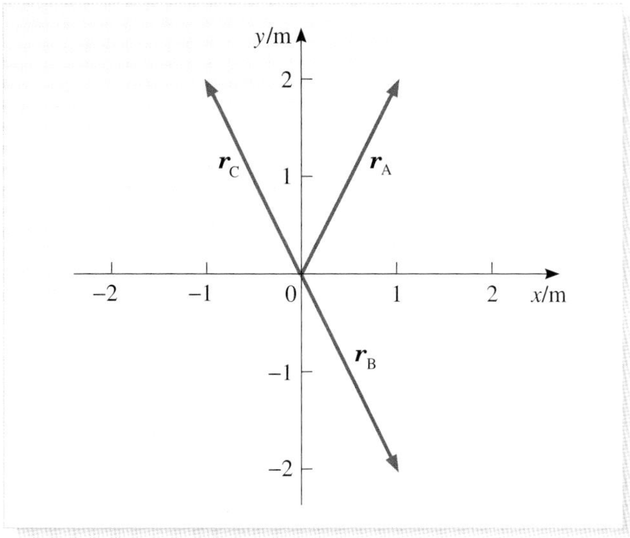

Figure 2.38 Diagram for Q2.4.

Q2.5 Let the displacement vector be $\boldsymbol{s} = (\Delta x, \Delta y)$.

(a) $s_x = \Delta x = 7\ \text{m} - 3\ \text{m} = 4\ \text{m}$

$s_y = \Delta y = -1\ \text{m} - 2\ \text{m} = -3\ \text{m}$.

(b) $s = \sqrt{(\Delta x)^2 + (\Delta y)^2} = \sqrt{4^2 + 3^2}\ \text{m} = 5\ \text{m}$.

(c) It is worth drawing a rough diagram showing $\boldsymbol{s}$, as in Figure 2.39. The figure also shows the same displacement $\boldsymbol{s}$ drawn from the origin to the point (4, −3). Remember that it doesn't matter where on the page you draw the displacement vector; only its magnitude and direction are important. The angle with the x-axis is given by $\theta = \arccos(0.8) = 36.9°$.

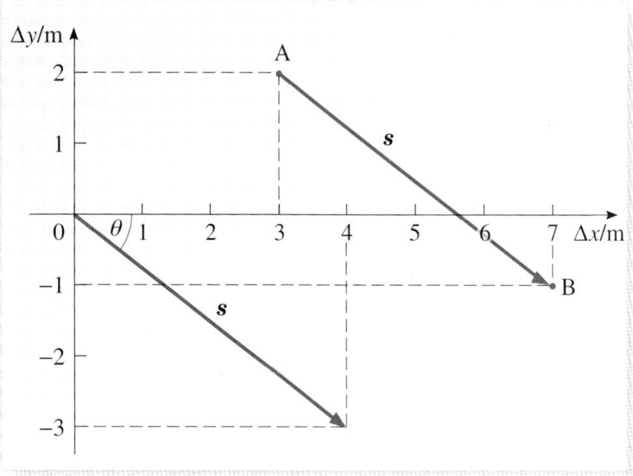

Figure 2.39 Diagram for Q2.5.

Q2.6 The components of the displacement vector of a particle from the origin are the *position coordinates* of the particle. Notice that it is important to distinguish between a *position* vector and a *displacement* vector. A position vector is represented by an arrow that always has its tail at the origin. A displacement vector does not, in general, have its tail tied to any particular point.

Q2.7 The components of the three displacement vectors are given by $\boldsymbol{a} = (7.7, 3.7)$ m, $\boldsymbol{b} = (2.2, -2.45)$ m and $\boldsymbol{c} = (9.9, 1.25)$ m, which are consistent with

$$c_x = a_x + b_x \quad \text{and} \quad c_y = a_y + b_y.$$

Q2.8 Figure 2.40a shows how $\boldsymbol{c} = \boldsymbol{a} + \boldsymbol{b}$, according to the triangle rule. We can see that in general the sum of the two lengths a and b is greater than the length c. The only way that we can have $c = a + b$, is if $\boldsymbol{a}$ and $\boldsymbol{b}$ are in the same direction, as in Figure 2.40b.

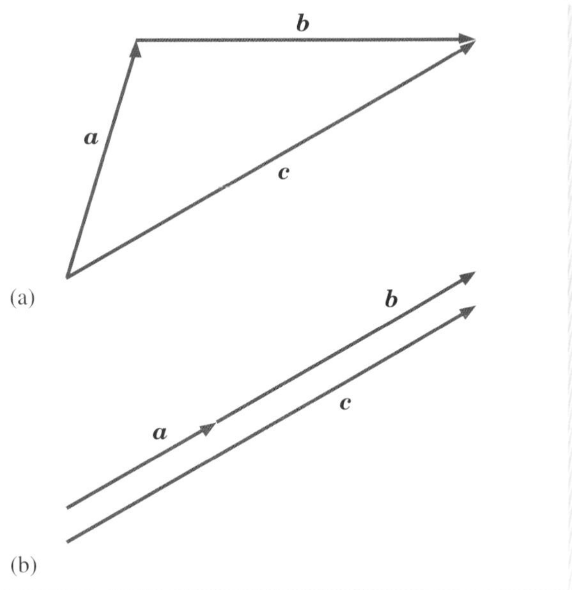

Figure 2.40 Diagram for Q2.8.

Q2.9 (a) We have $\boldsymbol{a} - 3\boldsymbol{b} = (3, 4)$ m $- 3(-1, 2)$ m $= (3 + 3, 4 - 3 \times 2)$ m $= (6, -2)$ m.

Hence the x-component of $\boldsymbol{a} - 3\boldsymbol{b}$ is 6 m and the y-component is -2 m.

(b) The vector $\boldsymbol{c}$ is given by $\boldsymbol{c} = (3, 4)\,\text{m} + (-1, 2)\,\text{m} = (2, 6)\,\text{m}$.

The lengths a, b, c are given by

$$a = \sqrt{3^2 + 4^2}\ \text{m} = 5\,\text{m}$$

$$b = \sqrt{1^2 + 2^2}\ \text{m} = \sqrt{5}\ \text{m} = 2.24\,\text{m}$$

$$c = \sqrt{2^2 + 6^2}\ \text{m} = \sqrt{40}\ \text{m} = 6.33\,\text{m}.$$

Hence $a + b = 5\,\text{m} + 2.24\,\text{m} = 7.24\,\text{m}$, which demonstrates that $a + b$ is not equal to c. This is as expected, since the vectors $\boldsymbol{a}$ and $\boldsymbol{b}$ are not in the same direction.

Q2.10 A displacement vector is fully characterized by its magnitude and direction. The magnitudes and directions of the two given displacements are certainly the same, so we can say that $\boldsymbol{a} = \boldsymbol{b}$ in this case. In addition the two vectors point in the same direction, so *in this particular case* we can also say that $|\boldsymbol{a} + \boldsymbol{b}| = a + b$.

Q2.11 Using the result for $\boldsymbol{b}$ in Q2.7, we have

$$\langle \boldsymbol{v} \rangle = \frac{\Delta \boldsymbol{r}}{\Delta t} = \frac{(2.2, -2.45)\,\text{m}}{(1.8 - 1.4)\,\text{s}}$$
$$= (5.5, -6.1)\,\text{m}\,\text{s}^{-1}.$$

Q2.12 Using the derivatives given in Table 1.6, we obtain

$$\boldsymbol{v} = \left(\frac{dx}{dt}, \frac{dy}{dt}\right) = (A,\ B - 2Ct).$$

On setting $t = 0.6$ s and substituting the values given for the constants, the expression for $\boldsymbol{v}$ reduces to

$$\boldsymbol{v} = (5.5,\ 9.5 - 2 \times 4.9 \times 0.6)\,\text{m}\,\text{s}^{-1}$$
$$= (5.5,\ 3.6)\,\text{m}\,\text{s}^{-1}.$$

Q2.13 (a) The speed of the ball is given by

$$v = \sqrt{v_x^2 + v_y^2} = \sqrt{(5.5)^2 + (3.6)^2}\ \text{m}\,\text{s}^{-1}$$
$$= 6.6\,\text{m}\,\text{s}^{-1}.$$

(b) Since $\cos\theta = \dfrac{v_x}{v} = \dfrac{5.5}{6.6} = 0.84$,

we have $\theta = \arccos(0.84) = 34°$.

Q2.14 Using the derivatives given in Table 1.6, we obtain

$$v = \left(\frac{dx}{dt}, \frac{dy}{dt}\right) = (A - 2Bt,\ A - 2Bt).$$

On setting $A = 9.5\,\text{m}\,\text{s}^{-1}$, $B = 4.9\,\text{m}\,\text{s}^{-2}$ and $t = 0$, we get

$$\boldsymbol{v} = (9.5, 9.5)\,\text{m}\,\text{s}^{-1}.$$

Q2.15 (a) The acceleration is given by

$$\boldsymbol{a} = \left(\frac{d^2x}{dt^2}, \frac{d^2y}{dt^2}\right) = \left(\frac{dv_x}{dt}, \frac{dv_y}{dt}\right).$$

Using the results given in Q2.12, we obtain

$$\boldsymbol{a} = (0, -2C) = (0, -2 \times 4.9)\,\text{m}\,\text{s}^{-2}$$
$$= (0, -9.8)\,\text{m}\,\text{s}^{-2}.$$

(b) $a = \sqrt{a_x^2 + a_y^2} = 9.8\,\text{m}\,\text{s}^{-2}$.

(c) The acceleration vector points in the direction of the negative y-axis. This implies that the acceleration is directed vertically downwards.

Q2.16 (a) The change in velocity is given by

$$\Delta \boldsymbol{v} = (0, 10)\,\text{m}\,\text{s}^{-1} - (10, 0)\,\text{m}\,\text{s}^{-1} = (-10, 10)\,\text{m}\,\text{s}^{-1}.$$

(b) Since the acceleration is constant in this case, we may say

$$\boldsymbol{a} = \frac{\Delta \boldsymbol{v}}{\Delta t} = (-10, 10)\,\text{m}\,\text{s}^{-2}.$$

(c) $a = \sqrt{a_x^2 + a_y^2} = \sqrt{100 + 100}\ \text{m}\,\text{s}^{-2}$
$$= 10\sqrt{2}\ \text{m}\,\text{s}^{-2} = 14.1\,\text{m}\,\text{s}^{-2}.$$

(d) If θ is the (anticlockwise) angle between the a_x-axis and $\boldsymbol{a}$, then

$$\cos\theta = \frac{a_x}{a} = \frac{-10}{10\sqrt{2}} = \frac{-1}{\sqrt{2}}$$

so $\theta = \arccos(-1/\sqrt{2}) = 135°$.

This result is confirmed by Figure 2.41, which also shows the direction of $\boldsymbol{a}$.

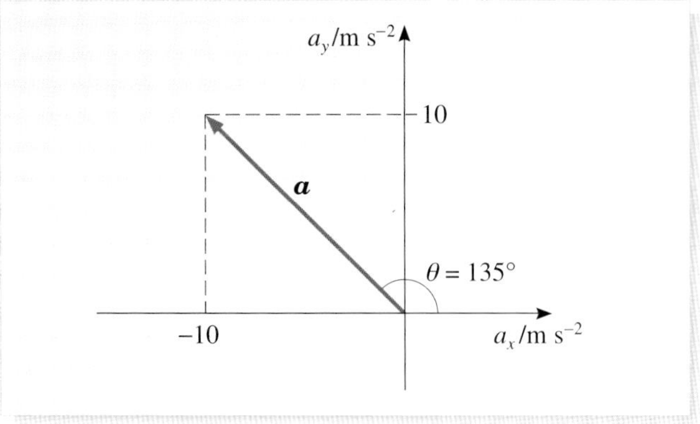

Figure 2.41 The acceleration vector in Q2.16.

(e) Both the initial and final speeds are 10 m s^{-1}. Therefore, the change in speed is equal to $\Delta v = 0$ m s^{-1}.

(f) $|\Delta\boldsymbol{v}| = \sqrt{(\Delta v_x)^2 + (\Delta v_y)^2} = \sqrt{100 + 100}$ m s^{-1}

i.e. $\Delta\boldsymbol{v} = 14.1$ m s^{-1}.

The symbol Δv means the change in v; that is, the change in the speed. It does *not* mean the magnitude of $\Delta\boldsymbol{v}$, which is written $|\Delta\boldsymbol{v}|$. In this question we have $\Delta v = 0$ m s^{-1}, but $|\Delta\boldsymbol{v}| = 14.1$ m s^{-1}.

Q2.17 The time of flight is equal to the duration of the vertical motion. This starts and ends with $s_y = 0$, and may be obtained from the equation

$$u_y t - \tfrac{1}{2} g t^2 = 0$$

i.e. $t(u_y - \tfrac{1}{2} g t) = 0$.

One solution is $t = 0$, which corresponds to the moment of launch. The other solution corresponds to the time of flight and is

$$t = \frac{2u_y}{g} = \frac{2u\sin\theta}{g} = \frac{2 \times 11.0 \sin 60°}{9.81}\,\text{s}$$
$$= 1.94\,\text{s}.$$

Q2.18 In Question 2.17, we found that the time of flight of the projectile is

$$t = \frac{2u_y}{g} = \frac{2u\sin\theta}{g} = 1.94\,\text{s}.$$

Since the horizontal velocity is constant throughout the time of flight, it follows that the final displacement of the ball from its launch position is

$$s_x = \frac{2u^2 \sin\theta\cos\theta}{g} = (1.94\,\text{s}) \times u\cos\theta.$$

It further follows that the range of the ball, $R = |s_x|$, is

$$R = (1.94\,\text{s}) \times (11.0\,\text{m s}^{-1}) \cos 60° = 10.7\,\text{m}.$$

If the ball is launched at 45° it will attain its maximum range, which for a launch speed of 11.0 m s^{-1} will be

$$R = \frac{u^2}{g} = \frac{(11.0\,\text{m s}^{-1})^2}{9.8\,\text{m s}^{-2}} = 12.3\,\text{m}.$$

Q2.19 From Equation 2.38, the required result is

$$s_y = \frac{11.0 \sin 60°}{11.0 \cos 60°} s_x - \frac{(9.81\,\text{m s}^{-2})}{2(11.0 \cos 60°)^2} s_x^2.$$
$$= 1.73 s_x - (0.162\,\text{m}^{-1}) s_x^2.$$

Q2.20 Multiplying out the parentheses in the given expression, we obtain

$$(x - 1)(x + 2) = x^2 + x - 2.$$

The right-hand side of this equation is of the form $ax^2 + bx + c$, provided $a = 1$, $b = 1$ and $c = -2$. It follows that the quadratic equation $x^2 + x - 2 = 0$, can be factorized as

$$(x - 1)(x + 2) = 0$$

and it is clear that this has the solutions $x = 1$ and $x = -2$.

Q2.21 The equation $x^2 - x - 2 = 0$ has the same form as Equation 2.40, with $a = 1$, $b = -1$, $c = -2$. Substituting these values into Equation 2.41 gives

$$x = \frac{-b \pm \sqrt{b^2 - 4ac}}{2a} = \frac{1 \pm \sqrt{1 + 4 \times 1 \times 2}}{2 \times 1}$$
$$= \frac{1 \pm 3}{2}.$$

Hence $x = 2$ and $x = -1$ are the solutions. We can check our answers by substitution into the original equation. For $x = 2$, we have $x^2 - x - 2 = 4 - 2 - 2 = 0$. For $x = -1$, we have $x^2 - x - 2 = 1 + 1 - 2 = 0$.

Q2.22 The equation $y^2 - 6y + 1 = 0$ has the same form as Equation 2.40, with $a = 1$, $b = -6$, $c = 1$ and x replaced by y. It follows that the solutions to the equation will be given by the quadratic equation formula (Equation 2.41) with x replaced by y and will be

$$y = \frac{-b \pm \sqrt{b^2 - 4ac}}{2a} = \frac{6 \pm \sqrt{36 - 4}}{2}$$
$$= \frac{6 \pm \sqrt{32}}{2} = \frac{6 \pm \sqrt{4 \times 8}}{2}$$
$$= \frac{6 \pm 2\sqrt{8}}{2} = 3 \pm \sqrt{8}$$
$$= 5.83 \text{ or } 0.172.$$

Hence $y = 5.83$ and $y = 0.172$ are the solutions. We can check our answers by substitution into the original equation. For $y = 5.83$, we have $y^2 - 6y + 1 = (5.83)^2 - 6 \times 5.83 + 1 = 0.009$. For $y = 0.172$, we have $y^2 - 6y + 1 = (0.172)^2 - 6 \times 0.172 + 1 = -0.002$. These results are consistent within the accuracy of our calculations.

Q2.23 The equation $s^2 - 4s + 4 = 0$ has the same form as Equation 2.40, with $a = 1$, $b = -4$, $c = 4$. Substituting into Equation 2.41 gives

$$s = \frac{-b \pm \sqrt{b^2 - 4ac}}{2a} = \frac{4 \pm \sqrt{16 - 4 \times 1 \times 4}}{2 \times 1}$$
$$= 2 \text{ or } 2.$$

Hence $s = 2$ and $s = 2$ are the solutions. (There are *two* solutions, although they both have the same value.) We can check our answers by substitution in the original equation: $s^2 - 4s + 4 = 4 - 8 + 4 = 0$.

Q2.24 Add 1 to both sides of the equation $x^2 - 1 = 0$ to obtain $x^2 = 1$. Hence $x = +1$ or $x = -1$. Alternatively write $x^2 - 1 = (x - 1)(x + 1) = 0$ or use Equation 2.41.

Q2.25 (a) The initial position of the particle is $\boldsymbol{r}_{\text{initial}} = (x_0, y_0) = (0, h)$ and the final position is $\boldsymbol{r}_{\text{final}} = (R, 0)$.

(b) The equations involving position coordinates (rather than displacements) that are equivalent to Equations 2.27 and 2.31 are

$$x - x_0 = u_x t \qquad (2.27a)$$

$$y - y_0 = u_y t - \tfrac{1}{2} g t^2. \qquad (2.31a)$$

In addition it is still the case that

$$v_x = u_x \qquad (2.26)$$

$$v_y = u_y - gt. \qquad (2.30)$$

(c) The time of flight, T, may be determined from Equation 2.31a by setting $y_0 = h$, $y = 0$ and $t = T$ to give

$$0 - h = u_y T - \tfrac{1}{2} g T^2$$

or $\tfrac{1}{2} g T^2 - u_y T - h = 0.$

A similar equation arose in the displacement-based discussion where we noted that it has only one positive solution

$$T = \frac{u_y + \sqrt{u_y^2 + 2gh}}{g}.$$

It then follows from Equation 2.27a that at the end of the flight, a time T after launch, $x = u_x T$, so

$$R = \left| \frac{u_x u_y + u_x \sqrt{u_y^2 + 2gh}}{g} \right|.$$

It follows from Equation 2.27a (with $x_0 = 0$) that $t = x/u_x$, which can be substituted into Equation 2.31a to give

$$y - h = u_y \left(\frac{x}{u_x} \right) - \tfrac{1}{2} g \left(\frac{x}{u_x} \right)^2$$

i.e. $$y = -\tfrac{1}{2} g \left(\frac{x}{u_x} \right)^2 + u_y \left(\frac{x}{u_x} \right) + h.$$

Which is the equation of a parabola in the xy-plane, since it is of the general form $y = ax^2 + bx + c$. The fact that in this case the constant term on the right-hand side ($c = h$) is not zero, means that the trajectory does not pass through the origin of coordinates, (0, 0), even though it does pass through the launch point $(0, h)$.

Q2.26 From Equation 2.38

$$-50\,\text{m} = \frac{(u \sin 30°)(200\,\text{m})}{u \cos 30°} - \frac{(9.81\,\text{m}\,\text{s}^{-2})(200\,\text{m})^2}{2u^2 \cos^2 30°}$$

i.e. $$-50\,\text{m} = 115\,\text{m} - \frac{2.62 \times 10^5\,\text{m}^3\,\text{s}^{-2}}{u^2}$$

Thus $$165\,\text{m} = \frac{2.62 \times 10^5\,\text{m}^3\,\text{s}^{-2}}{u^2},$$

so $$u^2 = \frac{2.62 \times 10^5\,\text{m}^3\,\text{s}^{-2}}{165.47\,\text{m}}$$

Since u must be positive, this implies that $u = 39.8\,\text{m}\,\text{s}^{-1}$.

Q2.27 The time of flight of the bullet will be the time that it takes to fall the 100 m to the sea, starting with an initial vertical component of velocity $u_y = 0$. It follows from Equation 2.46 (or Equation 2.48 if you prefer) that this time will be

$$T = \sqrt{\frac{2h}{g}} = \sqrt{\frac{200}{9.81}}\,\text{s} = 4.52\,\text{s}.$$

The horizontal displacement after this time of flight will be

$$s_x = u_x T = (300\,\text{m s}^{-1}) \times (4.52\,\text{s}) = 1.36 \times 10^3\,\text{m}$$

and, since this happens to be positive, it will also be the horizontal distance travelled by the bullet.

Q2.28 In the absence of air resistance the two bullets will hit the sea at the same time. The horizontal and vertical motions of a projectile are independent, so the horizontal motion of the bullet has no influence on its time of fall.

Q2.29 Using Equation 2.52,

$$r = \sqrt{x^2 + y^2 + z^2} = \sqrt{2^2 + 3^2 + 5^2}\,\text{m}$$
$$= \sqrt{38}\,\text{m} = 6.16\,\text{m}.$$

Q2.30 Substituting $\theta_x = \theta_y = 60°$ and $\theta_z = 45°$ and $r = 10\,\text{m}$ into Equation 2.53

$$x = r\cos\theta_x = r\cos 60° = 10 \times \tfrac{1}{2}\,\text{m} = 5\,\text{m}$$

$$y = r\cos\theta_y = r\cos 60° = 10 \times \tfrac{1}{2}\,\text{m} = 5\,\text{m}$$

$$z = r\cos\theta_z = r\cos 45° = 10 \times \tfrac{1}{\sqrt{2}}\,\text{m} = 7.07\,\text{m}.$$

Q2.31 (a) Letting $\theta = \theta_x = \theta_y = \theta_z$, we have

$$\cos^2\theta_x + \cos^2\theta_y + \cos^2\theta_z = 3\cos^2\theta = 1.$$

so $\cos\theta = \pm\frac{1}{\sqrt{3}}$.

Using a calculator gives $\arccos(1/\sqrt{3}) = 54.7°$ and $\arccos(-1/\sqrt{3}) = 125°$. Hence $\theta = 54.7°$ is the required answer. (There are other solutions to the equations $\cos\theta = \pm 1/\sqrt{3}$, but these do not lie between 0° and 90°.)

(b) $x = r\cos\theta_x = 10 \times \frac{1}{\sqrt{3}}\,\text{m} = 5.77\,\text{m}.$

The y- and z-components are also 5.77 m.

Q2.32 (a) $\boldsymbol{v} = \dfrac{d\boldsymbol{r}}{dt} = \dfrac{d}{dt}(at^2, bt, c) = (2at, b, 0).$

(b) The magnitude of v is

$$v = \sqrt{v_x^2 + v_y^2 + v_z^2} = \sqrt{(2at)^2 + b^2}$$
$$= \sqrt{4a^2t^2 + b^2}.$$

(c) $\boldsymbol{a} = \dfrac{d\boldsymbol{v}}{dt} = \dfrac{d}{dt}(2at, b, 0) = (2a, 0, 0).$

Q2.33 (a) Substituting $\boldsymbol{u} = (1, -10, 3)\,\text{m s}^{-1}$, $\boldsymbol{a} = (2, 7, -4)\,\text{m s}^{-2}$ and $t = 2\,\text{s}$

into $\boldsymbol{v} = \boldsymbol{u} + \boldsymbol{a}t$

gives $\boldsymbol{v} = (1, -10, 3)\,\text{m s}^{-1} + 2 \times (2, 7, -4)\,\text{m s}^{-1}$

$$= (1 + 2\times 2, -10 + 2\times 7, 3 - 2\times 4)\,\text{m s}^{-1}$$
$$= (5, 4, -5)\,\text{m s}^{-1}.$$

(b) The corresponding speed is given by

$$v = \sqrt{v_x^2 + v_y^2 + v_z^2}$$
$$= \sqrt{5^2 + 4^2 + (-5)^2}\,\text{m s}^{-1}$$
$$= \sqrt{66}\,\text{m s}^{-1} = 8.12\,\text{m s}^{-1}.$$

(c) Substituting the given values into

$$\boldsymbol{s} = \boldsymbol{u}t + \tfrac{1}{2}\boldsymbol{a}t^2,$$

gives $\boldsymbol{s} = 2(1, -10, 3)\,\text{m} + \dfrac{2^2}{2} \times (2, 7, -4)\,\text{m}$

$$= (2 + 4, -20 + 14, 6 - 8)\,\text{m}$$
$$= (6, -6, -2)\,\text{m}.$$

(d) $s = \sqrt{s_x^2 + s_y^2 + s_z^2} = \sqrt{76}\,\text{m} = 8.72\,\text{m}.$

Q2.34 (a) The fundamental equation for this problem is

$$\boldsymbol{s} = \boldsymbol{u}t + \tfrac{1}{2}\boldsymbol{a}t^2,$$

(i) To calculate the time of flight, we can use the y-component equation

$$s_y = u_y t + \tfrac{1}{2}a_y t^2$$

with $s_y = 0$ and $a_y = -g$, to obtain

$$0 = u_y t - \tfrac{1}{2}gt^2.$$

Dividing this equation by t (since we are not interested in the $t = 0$ solution) we find

$$t = \frac{2u_y}{g} = \frac{2(40.0\,\text{m s}^{-1})\sin 60°}{(9.81\,\text{m s}^{-2})}$$
$$= \frac{2 \times 40.0 \times \sqrt{3}}{9.81 \times 2}\,\text{s}$$
$$= 7.06\,\text{s}.$$

(Notice that we have used $\sin 60° = \sqrt{3}/2$, a useful exact relationship to remember.)

(ii) To find the x-displacement we use this value of t, together with $a_x = 0$, in the x-component equation to obtain

$$s_x = u_x t + \tfrac{1}{2}a_x t^2 = ut\cos 60°$$

$$= (40.0\,\mathrm{m\,s^{-1}}) \times (7.06\,\mathrm{s}) \times \tfrac{1}{2}$$

$$= 141\,\mathrm{m}.$$

We would therefore expect the golf ball to land at the position $\boldsymbol{r} = (141, 0, 0)\,\mathrm{m}$.

(b) A similar calculation gives the z-displacement, except that $u_z = 0\,\mathrm{m\,s^{-1}}$ and $a_z = 2\,\mathrm{m\,s^{-2}}$.

$$s_z = u_z t + \tfrac{1}{2}a_z t^2 = \tfrac{1}{2}a_z t^2$$

$$= \tfrac{1}{2} \times (2.00\,\mathrm{m\,s^{-2}}) \times (7.06\,\mathrm{s} - 2.00\,\mathrm{s})^2$$

$$= 25.6\,\mathrm{m}.$$

The position where the golf ball lands is therefore $\boldsymbol{r} = (141, 0, 25.6)\,\mathrm{m}$.

Q2.35 (a) $v = \sqrt{v_x + v_y} = \sqrt{3^2 + 4^2}\,\mathrm{m\,s^{-1}} = 5\,\mathrm{m\,s^{-1}}$.

(b) Since $\cos\theta = 3/5$, we have $\theta = \arccos(3/5) = 53.1°$.

Q2.36 (a) The (instantaneous) position vector of a moving particle is a vector that connects the origin of the coordinate system to the position of the particle at any particular time. It is defined in terms of the particle's position coordinates by

$$\boldsymbol{r}(t) = (x, y, z).$$

The (instantaneous) velocity of a particle is the rate of change of the particle's position at any time, and is given by

$$\boldsymbol{v}(t) = \frac{\mathrm{d}\boldsymbol{r}}{\mathrm{d}t}.$$

The (instantaneous) acceleration of a particle is the rate of change of the particle's velocity at any time, and is given by

$$\boldsymbol{a}(t) = \frac{\mathrm{d}\boldsymbol{v}}{\mathrm{d}t}.$$

(b) (i) Since $\boldsymbol{r}(t) = (At + 3Bt^2, 2At + 2Bt^2, 3At + Bt^2)$

$$r^2 = (At + 3Bt^2)^2 + (2At + 2Bt^2)^2 + (3At + Bt^2)^2$$

$$= A^2t^2(1^2 + 2^2 + 3^2) + 2ABt^3(3 + 4 + 3)$$

$$+ B^2t^4(1^2 + 2^2 + 3^2),$$

$$= 14A^2t^2 + 20ABt^3 + 14B^2t^4$$

then $r = \sqrt{14A^2t^2 + 20ABt^3 + 14B^2t^4}$.

(ii) Differentiating each of the components of the position vector (in accordance with the results given in Table 1.6)

$$\boldsymbol{v} = A(1, 2, 3) + 2Bt(3, 2, 1),$$

$$= (A + 6Bt, 2A + 4Bt, 3A + 2Bt).$$

The components of $\boldsymbol{v}$ are given by $v_x = A + 6Bt$, $v_y = 2A + 4Bt$, $v_z = 3A + 2Bt$.

(iii) We find $\boldsymbol{a} = (6B, 4B, 2B)$, so the components of $\boldsymbol{a}$ are $a_x = 6B$, $a_y = 4B$, $a_z = 2B$.

Q2.37 (a) One solution is $x = 0$. If x is non-zero, then we can divide all terms in the given equation by x, to obtain $x - 7 = 0$ and hence $x = 7$. The two solutions are therefore $x = 0$ and $x = 7$.

(b) The two solutions are given by

$$x = \pm\sqrt{7} = \pm 2.65.$$

(Never forget that a positive quantity has both a positive *and* a negative square root.)

(c) Substituting $a = 4$, $b = 4$ and $c = -3$ into the quadratic equation formula

$$x = \frac{-b \pm \sqrt{b^2 - 4ac}}{2a},$$

$$x = \frac{-4 \pm \sqrt{4^2 + 4 \times 4 \times 3}}{2 \times 4}$$

$$= \frac{-1 \pm 2}{2}.$$

Hence the two solutions are $x = 1/2$ and $x = -3/2$.

Q2.38 (a) A suitable system of coordinates is shown in Figure 2.42 where you can see that $\sin\theta = (10\,\mathrm{m})/(50\,\mathrm{m}) = 1/5$. In this system the launch point is at the origin, and the initial vertical component of velocity is given by

$$u_y = u\sin\theta = (25\,\mathrm{m\,s^{-1}}) \times \tfrac{1}{5}$$

$$= 5\,\mathrm{m\,s^{-1}}.$$

Since $v_y = 0$ at the maximum height, we can use the following equation of uniformly accelerated motion in the y-direction

$$v_y^2 = u_y^2 + 2a_y s_y$$

to get $$s_y = \frac{-u_y^2}{2a_y} = \frac{(5\,\mathrm{m\,s^{-1}})^2}{2 \times (9.81\,\mathrm{m\,s^{-2}})} = 1.3\,\mathrm{m}.$$

The maximum height above the ground is therefore $10\,\mathrm{m} + 1.3\,\mathrm{m} = 11.3\,\mathrm{m}$.

(b) We know that when the motor cyclist lands, $s_y = -10\,\mathrm{m}$. So, using another of the equations of uniformly accelerated motion,

$$s_y = u_y t + \tfrac{1}{2}a_y t^2$$

gives $-10\,\mathrm{m} = (5\,\mathrm{m\,s^{-1}})t + \tfrac{1}{2}(-9.8\,\mathrm{m\,s^{-2}})t^2$

which we can simplify to

$$(-4.9\,\mathrm{s^{-2}})t^2 + (5\,\mathrm{s^{-1}})t + 10 = 0.$$

This is a quadratic equation in t. According to Equation 2.41, its solutions are

$$t = \frac{-5 \pm \sqrt{5^2 + 4 \times 4.9 \times 10}}{-2 \times 4.9}\,\text{s}$$
$$= \frac{-5 \pm 14.9}{-9.8}\,\text{s}.$$

Since only the positive solution is meaningful in the current context, we have $t = 2.0\,\text{s}$.

(c) The horizontal component of the velocity on take-off is given by

$$u_x = u\cos\theta$$
$$= (25\,\text{m}\,\text{s}^{-1}) \times \frac{\sqrt{(50)^2 - (10)^2}}{50}$$
$$= 24.5\,\text{m}\,\text{s}^{-1}.$$

(Notice that we have used Pythagoras' theorem to obtain the horizontal length of the ramp that was not specified in the question.) The horizontal component of velocity remains unchanged throughout the motion. The vertical component of the velocity on landing is given by

$$v_y = u_y + a_y t.$$

so $v_y = (5\,\text{m}\,\text{s}^{-1}) + (-9.8\,\text{m}\,\text{s}^{-2}) \times (2\,\text{s}) = -14.6\,\text{m}\,\text{s}^{-1}$. (This is negative as expected, since the motor cyclist is descending.)

Hence, the landing speed is obtained as follows

$$v = \sqrt{v_x^2 + v_y^2}$$
$$= \sqrt{(24.5)^2 + (14.6)^2}\,\text{m}\,\text{s}^{-1}$$
$$= 28.5\,\text{m}\,\text{s}^{-1}.$$

(d) The horizontal displacement from the launch point to the landing point is

$$s_x = u_x t = 24.5 \times 2.0\,\text{m} = 49.0\,\text{m}.$$

Ten cars make a line of length $10 \times 4.50\,\text{m} = 45.0\,\text{m}$, which seems to imply that the motor cyclist can leap over 10 cars with a little over 4 m to spare. However, it would be foolhardy to attempt this feat, since

- Air resistance will slow the motor cyclist down during the flight.
- The motor cyclist will come down at an angle to the vertical, so might not clear the last car in the line, which will have a non-zero height.
- The motor cyclist is not really a particle, and allowance should be made for the non-zero size of the machine and its rider.

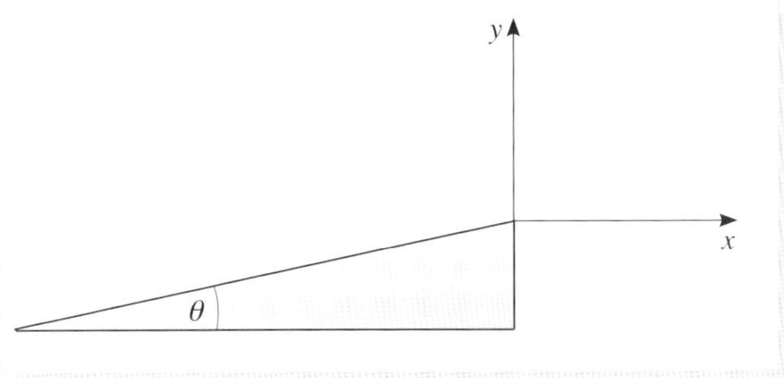

Figure 2.42 A set of coordinate axes for Q2.38.

Q2.39 (a) The equations showing how the components of velocity and displacement change with time are

$$v_x = u_x$$
$$v_y = u_y - gt$$
$$s_x = u_x t$$
$$s_y = u_y t - \tfrac{1}{2} g t^2.$$

In these equations: t is the time, g is the magnitude of the acceleration due to gravity, v_x is the x-component of the velocity at time t, v_y is the y-component of the velocity at time t, u_x is the x-component of the velocity at time $t = 0$, u_y is the y-component of the velocity at time $t = 0$, s_x and s_y are the x- and y-components of the displacement.

(b) An appropriate system of coordinates for the given situation is shown in Figure 2.43. The angle of launch will be determined by the ratio of the initial velocity components: $\tan\theta = u_y/u_x$. So our aim is to find suitable expressions for those components. The minimum horizontal component must be equal to the speed of the target, or the projectile will not keep up. Thus,

$$u_x = u_T.$$

The minimum vertical component of initial velocity will be that which just allows the projectile to attain the altitude of the target when $u_x = u_T$. Since the projectile is keeping pace with the target when $u_x = u_T$, this will ensure that the two meet. The projectile attains a maximum height h when $v_y = 0$. According to Equation 2.30 this occurs when $t = u_y/g$.

It therefore follows from Equation 2.31 (using $s_y = h$ at $t = u_y/g$) that

$$h = u_y\frac{u_y}{g} - \frac{1}{2}g\left(\frac{u_y}{g}\right)^2 = \frac{u_y^2}{2g}.$$

so $\quad u_y = \sqrt{2gh}$

Hence $\qquad \tan\theta = \dfrac{u_y}{u_x} = \dfrac{\sqrt{2gh}}{u_T}.$

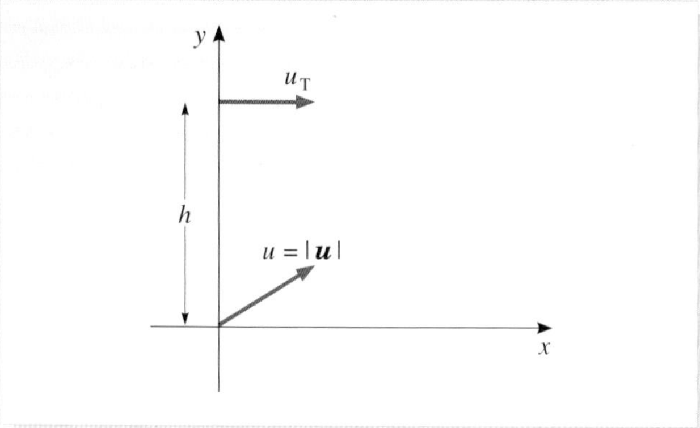

Figure 2.43 Suitable axes for Q2.39.

Q3.1 Since 2π rad = 360°, it follows that 1 rad = 360°/2π.

This is a simple calculator exercise leading to the result 57.295 779 5°.

Q3.2 (a) (i) $\dfrac{\pi}{2}\ \text{rad} = \dfrac{360^\circ}{2\pi} \times \dfrac{\pi}{2} = 90^\circ$;

(ii) $\dfrac{\pi}{3}\ \text{rad} = \dfrac{360^\circ}{2\pi} \times \dfrac{\pi}{3} = 60^\circ$;

(iii) $\dfrac{\pi}{4}\ \text{rad} = \dfrac{360^\circ}{2\pi} \times \dfrac{\pi}{4} = 45^\circ$;

(iv) $\pi\ \text{rad} = \dfrac{360^\circ}{2\pi} \times \pi = 180^\circ$.

(b) See Figure 3.43.

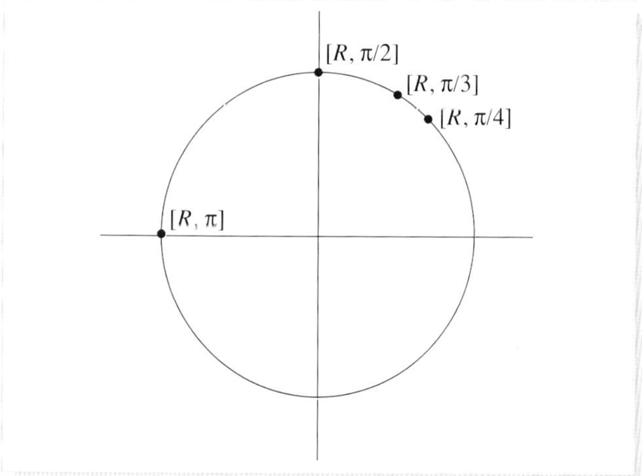

Figure 3.43 For Q3.2(b).

(c) The angular separation of the two points is $(\pi/2 - \pi/3)$ rad = $\pi/6$ rad. It follows that the required arc length is $s_{\text{arc}} = R\pi/6 = (2\ \text{m}) \times \pi/6 = 1.047\ \text{m}$.

Q3.3 (a) My calculator gives cos(630°) = 0.00 and sin(630°) = −1.00. (If you don't get these answers, you may have forgotten to set your calculator to degrees, rather than radians.) Hence the coordinates of the point are:

$$(x, y) = (0, -R) \text{ where } R = \sqrt{25\,\text{m}^2} = 5\,\text{m}.$$

(b) In this case, cos(−9.20 rad) = −0.975 and sin(−9.20 rad) = −0.223. Hence the coordinates are

$$(x, y) = (-5 \times 0.975\ \text{m}, -5 \times 0.223\ \text{m})$$
$$= (-4.87\ \text{m}, -1.11\ \text{m}).$$

Q3.4 This is really an investigation so there is no particular answer. However, an important point is that your calculator is a powerful tool for exploring the behaviour of the trigonometric functions. Make sure you know how to use it, and don't hesitate to explore the trigonometric functions with it.

Q3.5 (a) Expressing 27 days and 7 hours in terms of seconds we obtain $(27 \times 24 + 7) \times 3600$ s. Therefore, the angular speed is given by

$$\omega = \frac{2\pi}{(27 \times 24 + 7) \times 3600}\ \text{rad s}^{-1}$$
$$= 2.66 \times 10^{-6}\ \text{rad s}^{-1}.$$

(b) The angular speed will be (very nearly) constant in this case, so the time required is given by $\Delta t = \Delta\theta/\omega$. Since 1° is $(\pi/180)$ rad, so Δt, the time taken, is given by

$$\Delta t = \frac{\pi}{180}\ \text{rad} \times \frac{1}{2.66 \times 10^{-6}\ \text{rad s}^{-1}}$$
$$= 6.55 \times 10^3\ \text{s}.$$

Q3.6 Using Equation 3.22, we have $v = r\omega = (3.84 \times 10^8\ \text{m}) \times (2.66 \times 10^{-6}\ \text{rad s}^{-1}) = 1.02 \times 10^3\ \text{m s}^{-1}$.

Q3.7 Using Equation 3.21, we have

$$v = \frac{2\pi r}{T} = \frac{2\pi \times (6.38 \times 10^6\ \text{m})}{(24 \times 3600)\ \text{s}}$$
$$= 4.64 \times 10^2\ \text{m s}^{-1}.$$

Q3.8 From Equation 3.32, we have

$$a = \frac{v^2}{r} = \left(\frac{30 \times 10^3}{3600}\ \text{m s}^{-1}\right)^2 \times \frac{1}{40 \times 10^{-2}\ \text{m}}$$
$$= 174\ \text{m s}^{-2}.$$

Of course, this is not a complete answer; to specify an acceleration we also need to state its direction, which in this case is towards the centre of the wheel.

Q3.9 From Equation 3.31, we have

$$a = r\omega^2 = (3.84 \times 10^8\,\text{m}) \times (2.66 \times 10^{-6}\,\text{rad s}^{-1})^2 = 2.73 \times 10^{-3}\,\text{m s}^{-2}.$$

Q3.10 The angular speed of a satellite in geostationary orbit is

$$\omega = \frac{2\pi}{24 \times 60 \times 60}\,\text{rad s}^{-1} = 7.27 \times 10^{-5}\,\text{rad s}^{-1}.$$

If the radius of such an orbit is denoted R_C, then the centripetal acceleration of a satellite in orbit is of magnitude

$$a_C = R_C\omega^2.$$

But, according to Equation 3.33, the magnitude of the acceleration due to gravity at a distance R_C from the centre of the Earth is

$$a = \frac{4.00 \times 10^{14}\,\text{m}^3\,\text{s}^{-2}}{R_C^2}.$$

Hence $$R_C\omega^2 = \frac{4.00 \times 10^{14}\,\text{m}^3\,\text{s}^{-2}}{R_C^2}.$$

Rearranging this to isolate R_C we find

$$R_C = \sqrt[3]{\frac{4.00 \times 10^{14}\,\text{m}^3\,\text{s}^{-2}}{\omega^2}} = \sqrt[3]{\frac{4.00 \times 10^{14}\,\text{m}^3\,\text{s}^{-2}}{(7.27 \times 10^{-5}\,\text{rad s}^{-1})^2}} = 4.23 \times 10^7\,\text{m} = 42\,300\,\text{km}.$$

The symbol $\sqrt[3]{}$ used on the right-hand side above means the cube root of the enclosed quantity.

Q3.11 Using Equation 3.34 we obtain

$$f = \frac{1}{T} = \frac{1}{0.25\,\text{s}} = 4\,\text{Hz}.$$

Q3.12 Using Equation 3.36, and rearranging

$$\omega = \frac{2\pi}{T} = \frac{2\pi}{4\,\text{s}} = \frac{\pi}{2}\,\text{s}^{-1}.$$

Q3.13 When I shake hands, my hand moves up and down with an amplitude of about 80 mm and a period of about 0.5 s. The angular frequency of my handshake is therefore about $2\pi/(0.5\,\text{s}) = 4\pi\,\text{s}^{-1}$. If I choose to call the vertical direction the z-direction, and choose the midpoint of my handshake to be $z = 0$, then I can describe the motion of the centre of my hand by the equation,

$$z(t) = (80\,\text{mm})\sin[(4\pi\,\text{s}^{-1})t].$$

In writing this expression I have chosen to set the initial phase equal to zero, implying that I start to time my handshake from one of the moments when my hand is at the midpoint $z(0) = 0$ and initially moving in the direction of increasing z.

Q3.14 From Figures 3.10 and 3.11, we have, by inspection, $\sin(\theta + \pi/2) = \cos(\theta)$ and $\sin(\theta + \pi) = \cos(\theta + \pi/2)$. (Other expressions involving the cosine function are possible due to its symmetry and periodicity.)

Q3.15 (i) From Figure 3.28, we find the results given in Table 3.4.

Table 3.4 Table for the answer to part (i) of Q3.15.

θ/rad	Gradient from Figure 3.28a	$\cos(\theta)$
0	1	1
$\pi/2$	0	0
π	−1	−1
$3\pi/2$	0	0

These values are the same and are consistent with Equation 3.44.

(ii) From Figure 3.28, we also find the results given in Table 3.5.

Table 3.5 Table for the answer to part (ii) of Q3.15.

θ/rad	Gradient from Figure 3.28b	$\sin(\theta)$
0	0	0
$\pi/2$	−1	1
π	0	0
$3\pi/2$	1	−1

These values differ in sign and are therefore consistent with Equation 3.45.

Q3.16 Using the rule for the derivative of a constant times a function (see the bottom row of Table 1.6):

$$\frac{\text{d}}{\text{d}t}[A\sin(\omega t + \phi)] = A\frac{\text{d}}{\text{d}t}[\sin(\omega t + \phi)] = A\omega\cos(\omega t + \phi),$$

where we have used Equation 3.48. Similarly, for the second derivative

$$\frac{d^2}{dt^2}[A\sin(\omega t+\phi)] = \frac{d}{dt}\left(\frac{d}{dt}[A\sin(\omega t+\phi)]\right)$$
$$= \frac{d}{dt}\left(A\frac{d}{dt}[\sin(\omega t+\phi)]\right)$$
$$= \frac{d}{dt}[A\omega\cos(\omega t+\phi)]$$
$$= A\omega\frac{d}{dt}[\cos(\omega t+\phi)]$$
$$= A\omega^2[-\sin(\omega t+\phi)]$$
$$= -A\omega^2\sin(\omega t+\phi).$$

Q3.17 Equation 3.63 is the equation of s.h.m. with

$$\omega^2 = \frac{A\rho g}{M}.$$

It follows that the period of the drum's oscillations will be

$$T = \frac{2\pi}{\omega} = 2\pi\sqrt{\frac{M}{A\rho g}}.$$

Substituting the given values we find

$$A\rho g = \pi(0.35\,\text{m})^2 \times 1.025 \times 10^3\,\text{kg m}^{-3} \times 9.81\,\text{m s}^{-2}$$
$$= 3.87 \times 10^3\,\text{kg s}^{-2}.$$

so $T = 2\pi\sqrt{\dfrac{80\,\text{kg}}{3.87\times10^3\,\text{kg s}^{-2}}} = 0.90\,\text{s}.$

Q3.18 Dividing both sides of the equation by $16\,\text{m}^2$ we have

$$\frac{x^2}{16\,\text{m}^2} + \frac{4y^2}{16\,\text{m}^2} = \frac{x^2}{(4\,\text{m})^2} + \frac{y^2}{(2\,\text{m})^2} = 1.$$

Comparing this result with Equation 3.64, we deduce that $a = 4$ m and $b = 2$ m, and substituting these values into Equation 3.65 gives

$$e = \frac{1}{a}\sqrt{a^2 - b^2} = \frac{1}{4}\sqrt{16-4}$$
$$= \frac{\sqrt{3}}{2} = 0.866.$$

The foci are at $(\pm ae, 0) = (\pm 3.46\,\text{m}, 0)$.

Q3.19 Since, a is a length, it must satisfy $a \geq 0$ and therefore e cannot be negative. We can write Equation 3.65 as

$$e = \sqrt{1 - \frac{b^2}{a^2}} \tag{3.77}$$

and, in order to clarify this discussion, a graph of e against b^2a^{-2} is given in Figure 3.44. From Equation 3.77, we can see that we must have $b^2/a^2 \leq 1$ for e to be real. Hence, we find $e \geq 0$ which is the required lower limit. For the limiting case $e = 0$, Equation 3.64 reduces to $x^2 + y^2 = a^2$, which we recognize as the equation of a circle of radius a, centred on the origin. Therefore, a circle is merely a special case of an ellipse.

As b^2/a^2 gets smaller, e will increase, tending towards 1 for a^2 much greater than b^2 (written $a^2 >> b^2$). The value of e can only equal 1 if b^2/a^2 equals zero. Multiplying both sides of Equation 3.64 by b^2 and setting $b^2/a^2 = 0$, we can see that this limiting case corresponds to $y = \pm b$, that is, two straight lines, parallel to the x-axis and passing through $y = +b$ and $y = -b$. Since this case is not considered to be an ellipse, it is excluded by insisting that $e < 1$, rather than $e \leq 1$.

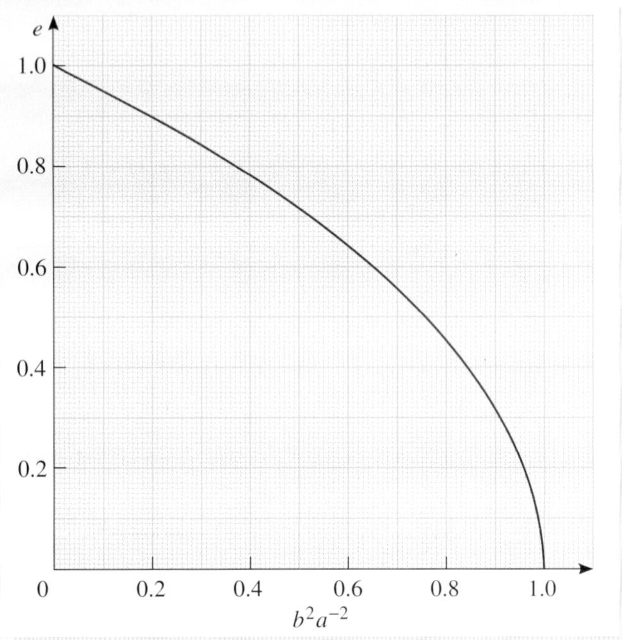

Figure 3.44 Graph for Q3.19.

Q3.20 Applying Kepler's third law to the two planets of this system, and remembering that the system is characterized by a different value of K (call it K_{51}) from the Solar System, we see that

$$K_{51} = \frac{(4.23\,\text{days}/365\,\text{days year}^{-1})^2}{(7.50\times10^6\,\text{km})^3}$$

$$\text{and } K_{51} = \frac{T^2}{(149.5\times10^6\,\text{km})^3}.$$

Note that we have taken the semimajor axis of the second planet (equal to that of the Earth) from Table 3.2, and converted the period of the first planet to years. It follows that

$$T = \left(\frac{4.23}{365}\right)\left(\frac{149.5}{7.50}\right)^{3/2} \text{year} = 1.03\,\text{year}.$$

Q3.21 Modifying Equation 3.76, we obtain

$$\begin{aligned} T_\text{H} &= T_\text{E}\left[\frac{1+(T_\text{S}/T_\text{E})^{2/3}}{2}\right]^{3/2} \\ &= \left[\frac{1+(29.5)^{2/3}}{2}\right]^{3/2} \text{years} \\ &= 12.11\,\text{years}. \end{aligned}$$

Hence the time of flight is $T_\text{H}/2 = 6.06$ years. (By exploiting the gravitational field of Jupiter, the *Voyager 1* spacecraft was actually able to make the trip in about 3 years.)

Q3.22 One complete orbit would correspond to 2π rad and therefore 2 days and 10 hours corresponds to

$$\frac{2\pi \times (2\times 24+10)}{(27\times 24+7)} = 5.56\times 10^{-1}\,\text{rad}.$$

In degrees, this is $\dfrac{5.56\times 10^{-1}\times 360°}{2\pi} = 31.9°$.

Q3.23 (a) Using Equation 3.20, we have

$$\begin{aligned} \omega &= \frac{2\pi}{T} \\ &= \frac{2\pi}{0.615\times 365\times 24\times 3600}\,\text{rad s}^{-1} \\ &= 3.24\times 10^{-7}\,\text{rad s}^{-1}. \end{aligned}$$

(b) From Equation 3.31, we obtain

$a = r\omega^2 = (1.08\times 10^{11}\,\text{m}) \times (3.24\times 10^{-7}\,\text{rad s}^{-1})^2$
$= 1.13\times 10^{-2}\,\text{m s}^{-2}$.

The acceleration is directed towards the Sun.

Q3.24 (a) Graphs of $\sin(\theta)$ and $\cos(\theta)$ are given in Figures 3.10 and 3.11.

(b) (i) $\theta = n\pi$, (ii) $\theta = (2n+1)\pi/2$, (iii) $\theta = (4n+1)\pi/2$, (iv) $\theta = (2n+1)\pi$.

Q3.25 Using Equation 3.49 to differentiate the given expression for $x(t)$, we obtain

$$\begin{aligned} \frac{\mathrm{d}x(t)}{\mathrm{d}t} &= C\frac{\mathrm{d}\cos(\omega t+\phi)}{\mathrm{d}t} \\ &= -C\omega\sin(\omega t+\phi). \end{aligned}$$

Equation 3.48 enables us to differentiate this expression, giving

$$\begin{aligned} \frac{\mathrm{d}^2x(t)}{\mathrm{d}t^2} &= -C\omega\frac{\mathrm{d}\sin(\omega t+\phi)}{\mathrm{d}t} \\ &= -C\omega^2\cos(\omega t+\phi) \\ &= -\omega^2 x(t). \end{aligned}$$

Q3.26 (a) Equation 3.34 gives $f = 1/T = 1/(0.541\,\text{s}) = 1.85\,\text{Hz}$.

From Equation 3.36, we obtain

$$\omega = \frac{2\pi}{T} = \frac{2\pi\,\text{rad}}{0.541\,\text{s}} = 11.6\,\text{rad s}^{-1}.$$

(b) The general solution to the simple harmonic oscillator equation is Equation 3.35

$$x(t) = A\sin(\omega t+\phi).$$

We are given that $A = 3.52$ m and $x(0) = 2.34$ m and we have shown in part (a) that $\omega = 11.6\,\text{rad s}^{-1}$. At time $t = 0$ s we have $\sin(\phi) = x(0)/A = (2.34\,\text{m})/(3.52\,\text{m}) = 0.665$ and hence $\phi = \arcsin(0.665) = 0.727$ rad.

The required expression is $x(t) = A\sin(\omega t+\phi)$ with $A = 3.52$ m, $\omega = 11.6\,\text{rad s}^{-1}$ and $\phi = 0.727$ rad, i.e. $x(t) = (3.52\,\text{m})\sin(11.6t/\text{s} + 0.727)$.

Q3.27 Dividing $x^2 + 9y^2 = 9\,\text{m}^2$ by $9\,\text{m}^2$ gives

$$\frac{x^2}{(3\,\text{m})^2} + \frac{y^2}{(1\,\text{m})^2} = 1.$$

Comparing this result with Equation 3.64, gives $a = 3$ m and $b = 1$ m, where a and b are the semimajor and semiminor axes, respectively. From Equation 3.65, we find that

$$\begin{aligned} e &= \frac{1}{a}\sqrt{a^2-b^2} \\ &= \frac{1}{3}\sqrt{3^2-1^2} = \frac{2\sqrt{2}}{3}. \end{aligned}$$

The foci are $(\pm ae, 0) = (\pm 2\sqrt{2}, 0)$ m.

Q3.28 Let $R_\text{max} = 10^{10}$ km and $R_\text{min} = 10^8$ km. From Figure 3.34 you can see that $R_\text{max} = a(1+e)$ and $R_\text{min} = a(1-e)$. Hence

$$\begin{aligned} a &= \frac{R_\text{max}+R_\text{min}}{2} \\ &= \frac{10^{10}+10^8}{2}\,\text{km} \\ &= 5.05\times 10^9\,\text{km}. \end{aligned}$$

We also know that $e = \frac{1}{a}\sqrt{a^2 - b^2}$

and therefore

$$\begin{aligned} b &= \sqrt{a^2 - e^2a^2} \\ &= \sqrt{a(1+e) \times a(1-e)} \\ &= \sqrt{R_{max}R_{min}} \\ &= \sqrt{10^{10} \times 10^8}\ \text{km} = 10^9\ \text{km}. \end{aligned}$$

Q3.29 From Kepler's third law we would expect that the data obey $T = Ka^{3/2}$. However, the corresponding graph (Figure 3.45) is not a straight line. Consequently these (fictional) data do not support Kepler's law.

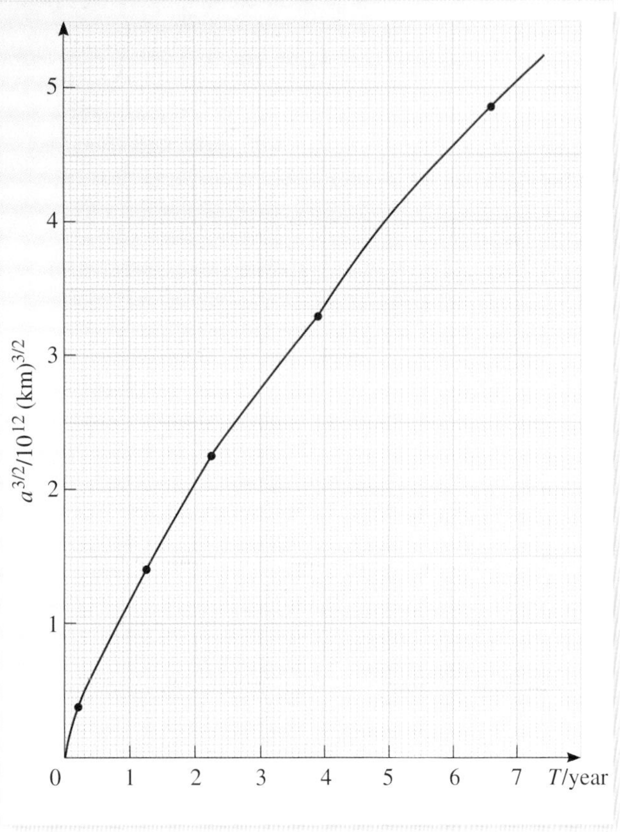

Figure 3.45 Graph for Q3.29.

(You will learn a better method of carrying out such tests, involving logarithms, later in the course.)

Q4.1 The position of P is –20 m. The displacement of P from Q is position of P minus position of Q = –20 m – 15 m = –35 m.

Q4.2 The velocity is $v_x = -35\ \text{m}/5\ \text{s} = -7\ \text{m s}^{-1}$.

Q4.3 Vector $2\boldsymbol{a}$ has magnitude 10 m and the same direction as $\boldsymbol{a}$, i.e. vertically upwards.

$-\boldsymbol{a}$ means $-1\boldsymbol{a}$; it has magnitude 5 m and points vertically downwards. The minus sign reverses the direction of the vector.

$-0.1\boldsymbol{a}$ is a vector of magnitude 0.5 m pointing vertically downwards.

Q4.4 The vector $\boldsymbol{b} + \boldsymbol{c}$ points north-east by the triangle rule, while vector $\boldsymbol{b} - \boldsymbol{c} = \boldsymbol{b} + (-\boldsymbol{c})$ points south-east. See Figure 4.9.

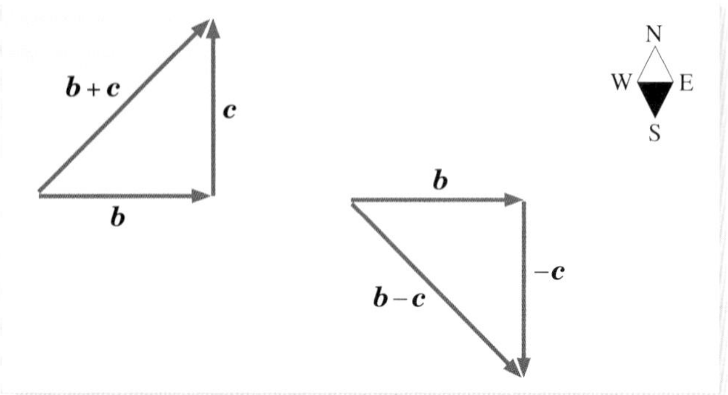

Figure 4.9 For use with Q4.4.

Q4.5 $a_x = 5\ \text{m} - (-2\ \text{m}) = 7\ \text{m}$, $a_y = 1\ \text{m} - 5\ \text{m} = -4\text{m}$. $b_x = (3\ \text{m})\cos(60°) = (3/2)\ \text{m}$, $b_y = (3\ \text{m})\sin(60°) = (3\sqrt{3}/2)\ \text{m}$. $c_x = 3\ \text{m}$, $c_y = -6\ \text{m}$.

Q4.6 (a) The magnitude of vector $\boldsymbol{c}$ is $|(3, -6)|$ m $= \sqrt{3^2 + (-6)^2}\ \text{m} = \sqrt{45}\ \text{m} = 3\sqrt{5}\ \text{m}$.

(b) The magnitude of $\boldsymbol{u}$ is $u = |\boldsymbol{u}| = \sqrt{2}\sqrt{1^2 + 1^2}\ \text{m} = \sqrt{2} \times \sqrt{2}\ \text{m} = 2\ \text{m}$. Alternatively write $\boldsymbol{u} = (\sqrt{2}, \sqrt{2})\ \text{m}$, and so $u = \sqrt{2 + 2}\ \text{m} = 2\ \text{m}$.

Q4.7 The acceleration of a projectile is always of magnitude g and directed vertically downwards.

Q4.8 The velocity is directed along the tangent to the circular path at P, in the direction from P towards R for anticlockwise motion. The centripetal acceleration is always directed towards the centre of the circle, i.e. in the direction from P towards O.

Q4.9 (a) The amplitude is $A = 2$ m. (b) It takes a quarter-period to move from the midpoint to one endpoint. Hence the period is 4 s.

Q4.10 (a) The speed is largest at the midpoint, i.e. at the origin in this case. (b) The acceleration is zero at the midpoint, i.e. at the origin.

Q4.11 The period is $T = 2$ s and so the angular frequency is $\omega = 2\pi/T = \pi\,\text{s}^{-1}$. The displacement–time graph would have to be shifted to the right by 1 s to coincide with the sine curve. This represents half a period, so the initial phase is $\phi = 2\pi \times 1/2 = \pi$.

Q4.12 (a) The velocity is defined to be the derivative

$$\boldsymbol{v} = \frac{\mathrm{d}\boldsymbol{r}}{\mathrm{d}t}.$$

(b) The acceleration is defined to be the derivative

$$\boldsymbol{a} = \frac{\mathrm{d}\boldsymbol{v}}{\mathrm{d}t} = \frac{\mathrm{d}^2\boldsymbol{r}}{\mathrm{d}t^2}.$$

Q4.13 A planet moves fastest when it is closest to the Sun.

Q4.14 The conversion is defined by 2π radians = 360°. Hence 40° = 40° × (2π radians/360°) = $2\pi/9$ radians, and π radians = π radians × (360°/2π radians) = 180°.

Q4.15 $s_{\text{arc}} = (\pi/2) \times 5\,\text{m} = (5\pi/2)\,\text{m}$.

Q4.16 The tangent line at 23 s is shown on the graph. Its gradient is $\dfrac{(44-24)\,\text{m}}{(40-10)\,\text{s}} = \dfrac{2}{3}\,\text{m}\,\text{s}^{-1}$. This is the velocity of the particle at $t = 23$ s. (The positive sign indicates the direction in this one-dimensional case.)

Q4.17 The particle is at $x = 6$ m at $t = 0$ s. The particle passes through the origin at $t = -4$ s.

Q4.18 The tangent line is shown at $t = 35$ s. Its gradient is $\dfrac{(18-45)\,\text{m}\,\text{s}^{-1}}{(50-20)\,\text{s}} = \dfrac{-27\,\text{m}\,\text{s}^{-1}}{30\,\text{s}} = -0.9\,\text{m}\,\text{s}^{-2}$. This is the acceleration of the particle at $t = 35$ s. (The sign indicates the direction.)

Q4.19 The speed of the particle at $t = 0$ s is 42 m s^{-1}. The particle is at rest at $t = 50$ s.

Q4.20 The three equations are given in factored form. The solutions are: (a) $x = 3$ and $x = -2$; (b) $x = 1$ and $x = 1$. (The two solutions are the same.) (c) $x = 0$ and $x = -5$.

Q4.21 (i) If b changes to $2b$, then the square root in the numerator (top part of the fraction) changes to $\sqrt{2b} = \sqrt{2}\sqrt{b}$ and so a increases by a factor of $\sqrt{2}$. (ii) If c triples in value then c^2 in the denominator (bottom part of the fraction) becomes $(3c)^2 = 9c^2$, and so a decreases by a factor of 9. (iii) If b increases by a factor of 4 then $\sqrt{b}$ in the numerator increases by a factor of $\sqrt{4} = 2$. To compensate for this the value of c^2 in the denominator would have to increase by a factor of 2. This means c would have to increase by a factor of $\sqrt{2}$.

Q4.22 The position at $t = 0$ is $x(0) = C$.

Its velocity and acceleration at $t = 0$ are

$$v_x(0) = \frac{\mathrm{d}x}{\mathrm{d}t}(0) = -B, \text{ and } a_x(0) = \frac{\mathrm{d}^2x}{\mathrm{d}t^2}(0) = 2A.$$

Acknowledgements

Grateful acknowledgement is made to the following sources for permission to reproduce material in this book:

Fig. 1.1 ZARM Droptower of Bremen at the University of Bremen; *Fig. 1.2* LERC/NASA; *Fig. 1.3* Alton Towers; *Fig. 1.23* Paolo Fioratti/Oxford Scientific Films (*swift in flight*), Keren Su/Oxford Scientific Films (*growth of a bamboo shoot*), Spectrum Colour Library (*growth of a child*); *Fig. 1.25* Science Photo Library (*explosion*), Popperfoto/Reuters (*surface-to-air missile*), NASA/Sci Mus/Sci & Soc Pic Lib (*highest rocket acceleration*), NASA/OSF (*lunar gravity*); *Fig. 1.36* ZARM Droptower of Bremen at the University of Bremen; *Fig. 2.1* Allsport; *Fig. 2.20* Allsport (*golfer*), OSF (*archer*); *Fig. 2.21* Mary Evans; *Fig. 2.22* British Library; *Fig. 2.27a* Salamander Picture Library; *Fig. 3.1 Top* Matra Marconi Space; *Fig. 3.1 Middle* Photo aerospatiale; *Fig. 3.1 Bottom* Matra Marconi Space; *Fig. 3.3* Picture courtesy of British Telecom; *Fig. 3.4b* Permission granted by Readers Digest Ltd — AA Book of the Car © 1976; *Fig. 3.36a* 1980 UKATC Royal Observatory Edinburgh; *Fig. 3.36b* Science & Society Picture Library; *Fig. 3.41* Armagh Planetarium.

Index

Entries and page numbers in **bold type** refer to key words which are printed in **bold** in the text and which are defined in the Glossary.